Battery Management Systems of Electric and Hybrid Electric Vehicles

Battery Management Systems of Electric and Hybrid Electric Vehicles

Editor

Nicolae Tudoroiu

MDPI • Basel • Beijing • Wuhan • Barcelona • Belgrade • Manchester • Tokyo • Cluj • Tianjin

Editor
Nicolae Tudoroiu
John Abbott College
Canada

Editorial Office
MDPI
St. Alban-Anlage 66
4052 Basel, Switzerland

This is a reprint of articles from the Special Issue published online in the open access journal *Batteries* (ISSN 2313-0105) (available at: https://www.mdpi.com/journal/batteries/special_issues/ Battery_Management_Systems_Electric_Hybrid_Electric_Vehicles).

For citation purposes, cite each article independently as indicated on the article page online and as indicated below:

LastName, A.A.; LastName, B.B.; LastName, C.C. Article Title. *Journal Name* **Year**, *Volume Number*, Page Range.

ISBN 978-3-0365-1060-6 (Hbk)
ISBN 978-3-0365-1061-3 (PDF)

© 2021 by the authors. Articles in this book are Open Access and distributed under the Creative Commons Attribution (CC BY) license, which allows users to download, copy and build upon published articles, as long as the author and publisher are properly credited, which ensures maximum dissemination and a wider impact of our publications.
The book as a whole is distributed by MDPI under the terms and conditions of the Creative Commons license CC BY-NC-ND.

Contents

About the Editor vii

Roxana-Elena Tudoroiu, Mohammed Zaheeruddin, Nicolae Tudoroiu and Sorin-Mihai Radu
SOC Estimation of a Rechargeable Li-Ion Battery Used in Fuel-Cell Hybrid Electric Vehicles—Comparative Study of Accuracy and Robustness Performance Based on Statistical Criteria. Part I: Equivalent Models
Reprinted from: *Batteries* **2020**, *6*, 42, doi:10.3390/batteries6030042 1

Roxana-Elena Tudoroiu, Mohammed Zaheeruddin, Nicolae Tudoroiu and Sorin-Mihai Radu
SOC Estimation of a Rechargeable Li-Ion Battery Used in Fuel Cell Hybrid Electric Vehicles—Comparative Study of Accuracy and Robustness Performance Based on Statistical Criteria. Part II: SOC Estimators
Reprinted from: *Batteries* **2020**, *6*, 41, doi:10.3390/batteries6030041 39

Fabio Cignini, Antonino Genovese, Fernando Ortenzi, Adriano Alessandrini, Lorenzo Berzi, Luca Pugi and Riccardo Barbieri
Experimental Data Comparison of an Electric Minibus Equipped with Different Energy Storage Systems
Reprinted from: *Batteries* **2020**, *6*, 26, doi:10.3390/batteries6020026 75

Alexander Fill, Tobias Mader, Tobias Schmidt, Raphael Llorente and Kai Peter Birke
Measuring Test Bench with Adjustable Thermal Connection of Cells to Their Neighbors and a New Model Approach for Parallel-Connected Cells
Reprinted from: *Batteries* **2020**, *6*, 2, doi:10.3390/batteries6010002 91

Jan Meyer, Antonio Nedjalkov, Elke Pichler, Christian Kelb and Wolfgang Schade
Development of a Polymeric Arrayed Waveguide Grating Interrogator for Fast and Precise Lithium-Ion Battery Status Monitoring
Reprinted from: *Batteries* **2019**, *5*, 66, doi:10.3390/batteries5040066 107

Manh-Kien Tran and Michael Fowler
Sensor Fault Detection and Isolation for Degrading Lithium-Ion Batteries in Electric Vehicles Using Parameter Estimation with Recursive Least Squares
Reprinted from: *Batteries* **2020**, *6*, 1, doi:10.3390/batteries6010001 121

About the Editor

Nicolae Tudoroiu In 1976 he received his B.Eng. degree in Electrical and Computer Engineering from the University of Craiova, Romania, and in 1981 received his B.Sc. in Mathematics from the West University of Timisoara, Romania. In 1990 obtained his Ph.D. degree in Automation in Romania and in, 2001, he also received a Ph.D. degree in Electrical and Computer Engineering, from Concordia University, Montreal, Canada. During the period 1979–1994, he joined the departments of Automation and Control of the University of Craiova, Romania (1979–1990), and West University "Politehnica" Timisoara, Romania (1991–1994), as an assistant and associate professor, respectively. In the period 2001–present he joined as tenure professor the Engineering Technologies department of John Abbott College, Sainte-Anne-de-Bellevue, Canada. He presently serves on the editorial board as Guest editor of the International Journal *'Advance in Science, Technology and Engineering Systems'* (*ASTESJ*). His interest in academic research includes system modelling and identification, process control, state and parameter estimation techniques, fault detection and isolation, neural networks, adaptive and optimal control systems.

Article

SOC Estimation of a Rechargeable Li-Ion Battery Used in Fuel-Cell Hybrid Electric Vehicles—Comparative Study of Accuracy and Robustness Performance Based on Statistical Criteria. Part I: Equivalent Models

Roxana-Elena Tudoroiu [1], Mohammed Zaheeruddin [2], Nicolae Tudoroiu [3,*] and Sorin-Mihai Radu [4]

1. Department of Mathematics and Informatics, University of Petrosani, 332006 Petrosani, Romania; tudelena@mail.com
2. Department of Building, Civil and Environmental Engineering, University Concordia from Montreal, Montreal, QC H3G 1M8, Canada; zaheer@encs.concordia.ca
3. Department of Engineering Technologies, John Abbott College, Saint-Anne-de-Bellevue, QC H9X 3L9, Canada
4. Department of Electrical and Power Engineering, University of Petrosani, 332006 Petrosani, Romania; sorin_mihai_radu@yahoo.com
* Correspondence: ntudoroiu@gmail.com or nicolae.tudoroiu@johnabbott.qc.ca; Tel.: +1-514-966-5637

Received: 15 July 2020; Accepted: 13 August 2020; Published: 14 August 2020

Abstract: Battery state of charge (SOC) accuracy plays a vital role in a hybrid electric vehicle (HEV), as it ensures battery safety in a harsh operating environment, prolongs life, lowers the cost of energy consumption, and improves driving mileage. Therefore, accurate SOC battery estimation is the central idea of the approach in this research, which is of great interest to readers and increases the value of its application. Moreover, an accurate SOC battery estimate relies on the accuracy of the battery model parameters and its capacity. Thus, the purpose of this paper is to design, implement and analyze the SOC estimation accuracy of two battery models, which capture the dynamics of a rechargeable SAFT Li-ion battery. The first is a resistor capacitor (RC) equivalent circuit model, and the second is a generic Simscape model. The model validation is based on the generation and evaluation of the SOC residual error. The SOC reference value required for the calculation of residual errors is the value estimated by an ADVISOR 3.2 simulator, one of the software tools most used in automotive applications. Both battery models are of real interest as a valuable support for SOC battery estimation by using three model based Kalman state estimators developed in Part 2. MATLAB simulations results prove the effectiveness of both models and reveal an excellent accuracy.

Keywords: SAFT lithium-ion battery; Simscape model; 3RC ECM Li-ion battery model; state of charge; ADVISOR estimate

1. Introduction

1.1. Literature Review

Nowadays the new technologies applied in batteries manufacturing industry "often demand more compact, higher capacity, safe and rechargeable batteries" [1]. The batteries vary by different chemistries and "generate the basic cell voltages typically in the 1.0 to 3.6 V range" [1]. The required voltages and the currents of a battery pack are obtained by adding up the number of the cells in a series connection to increase the voltage and parallel connection to enhance the current. It is important to note that when driving on the road, electric car batteries "need to be recharged relatively quickly", which is one of the key requirements for a Li-ion battery [2]. The maximum power required by a Li-ion battery pack for charging is calculated by using the following formula [2]:

$$P_{max} = V_{bat,max} \times N_{cells} \times I_{bat,max} \tag{1}$$

where $V_{bat,max}$ is the maximum terminal voltage of the cell, N_{cells} is the number of cells in the pack and $I_{bat,max}$ is the maximum charging current allowed per cell.

In contrast to other electric vehicles, "the fuel cell electric vehicles (FCEVs) produce, cleanly and efficiently, electricity using the chemical energy of a fuel cell powered by hydrogen, rather than drawing electricity from only a battery" [3,4]. A hybrid FCEV (HFCEV) can be designed "with plug-in capabilities to charge the battery", since "most HFCEVs today use the battery for recapturing braking energy, providing extra power during short acceleration events, and to smooth out the power delivered from the fuel cell" [3]. Compared to "conventional internal combustion engine vehicles", the HFCEVs "are more efficient and produce no harmful tailpipe emissions" [3].

Fuel cells (FC) "work like batteries, but they do not run down or need recharging; they produce electricity and heat as long as fuel is supplied" [4]. Among "the most common types of fuel cell for vehicle applications is the polymer electrolyte membrane (PEM) [3,4]. The most recent HFCEVs "are equipped with advanced technologies to increase efficiency, such as regenerative braking systems, which capture the energy lost during braking and store it in a battery" [3]. In the case study of a small hybrid electric vehicle (HEV) car (HEV-SMCAR), the fuel cell battery is "designed to meet the average load power, while batteries and supercapacitors provide extra power during transients and overload" [5]. This reduces drastically "the size of the fuel cell system" and also "improve the dynamic response of hybrid power system" [5]. As a new improvement of HEV performance it is more appropriate to consider a "hybridization of the on-board energy source, i.e., to combine the Li-ion battery, and energy source, with a component that is more power dense" [6].

The supercapacitor (SC) is used in this combination, since it is "able to provide high power for short periods of time without damaging their internal structure" and also works for a long life-cycle with a high efficiency, which exceeds the Li-ion battery performance [6–10]. Also, the SC keeps the discharging Li-ion battery current within battery limits given in specifications, such that to "extend the Li-ion battery life cycle by compensating "the high current of the load" [6]. To reach this objective an energy management system (EMS) is required [5–11]. In [5,6], the EMS is conceived as an algorithmic procedure for developing five EMS techniques, to optimize the hydrogen consumption, and to assure a high overall system efficiency, as well as a long-life cycle. In [7,11] a detailed diagram of an EMS system is presented, for an FC, UC, and Li-ion battery hybrid energy storage, to rationalize both power density and energy density, which can be adapted such that to be useful for a small hybrid electric car (SMCAR) proposed in our case study. To simplify the Simulink diagram of the EMS, the authors use Simscape components such as Li-ion battery, supercapacitor, and FCPEM [8]. The battery state of charge (SOC) is an essential internal parameter of the battery and SC/UC that is under observation constantly by a battery management system (BMS) to "prevent hazardous situations and to improve battery and SC/UC performance" [12–14]. Typically, for calculation, SOC is "tracking according to the discharging current" [14–16].

1.2. Li-Ion Battery Models Reported in the Literature—Brief Presentation

In the absence of a measurement sensor, the SOC cannot be measured directly, thus its estimation using Kalman filter estimation techniques is required [17–29], a topic that is detailed in Part 2 [30]. Furthermore, an accurate Li-ion battery model is essential in SOC estimation of the model-based BMS in electric vehicles (EVs)/HEVs. A complete analysis of the current state of the SOC estimation of the Li-ion battery for EVs is presented in [28]. In the paper it is stated that for EV/HEVs battery systems, "an accurate SOC can prevent battery discharge and charging, thus ensuring the safety of the battery system, more efficiently using limited energy and extending battery life". It can also "support the accurate calculation of the driving range of the vehicle, provide a better discharge or charging strategy, improve the efficiency of other energy sources and make balancing strategies work more efficiently". The same research paper [28] emphasizes that the design of the Li-ion battery model and the real-time implementation of adequate SOC estimation in EV/HEVs applications is a challenging task due to the "complexity of electrochemical reactions and performance degradation caused by various factors". Design and implementation in real time of stable, accurate and robust SOC estimation algorithms encounter some critical issues, such as hysteresis and the flat aspect of the open-circuit controlled voltage (OCV) = f(SOC) characteristic curve, Li-ion battery model, ageing, choice of estimation algorithm and imbalance cell [28]. Therefore, these issues require a comprehensive analysis to consider their impact on solving the correct and accurate battery SOC estimation. At the end of this section, a brief review is given of some linear and non-linear analytic battery models of different chemistries reported in the literature well-suited for "battery design, performance estimation, prediction for real-time power management, and circuit simulation", such as is done in [17,18,28]. These models can be categorized into five categories: electrochemical models, computational intelligence-based models, analytical models, stochastic models, and electrical circuit models, as is mentioned [17].

1.2.1. Electrochemical Battery Models

The electrochemical models or distributed physics-based models excel by their accuracy concerning the prediction of battery terminal output voltage, achieved by these models, but they require detailed knowledge of the battery chemical processes, which makes them difficult to configure [21,29].

Furthermore, these models can capture the electrochemical reactions using partial differential equations (PDE) "that links physical parameters to internal electrochemical dynamics of the battery cell allowing trade off analysis and high accuracy", as is stated in [19]. A well-known early model with a high accuracy of 2% was originally developed by Doyle, Fuller and Newman, as mentioned in [3]. A big advantage of these models is that the PDE equations deal with numerous unknown parameters, are more complicated computationally expensive, as well as they are almost impractical in real-time BMS applications.

1.2.2. Computational Intelligence-Based Battery Models

The second category of models are the computational intelligence-based models. These models describe the non-linear relationships between SOC, battery terminal voltage, input current, and battery internal temperature, and specifically use artificial neural network (ANN)-based models and support vector regression models, and mixed models. Also, mixed models have been used to estimate the battery non-linear behaviors. An interesting "recurrent neural network (RNN) has been used to build an SOC observer estimator and battery terminal voltage estimator", as is mentioned in [29].

1.2.3. Analytical Battery Models

The third category of models, namely the analytical models are simplified version of electrochemical models, such as those based on the Pucker's law, the kinetic battery models, and the diffusion models, as are detailed in [29]. These models could capture nonlinear capacity effects and to predict runtime of the batteries with reduced order of equations and perform well for battery SOC tracking and runtime prediction under specific discharge profiles [29].

1.2.4. Stochastic Battery Models

The fourth category describe the stochastic models that emphasize on "modeling recovery effect and describes battery behavior as a Markov process with probabilities in terms of parameters that are related to the physical characteristics of an electrochemical cell" [29]. They give a "good qualitative description for the behavior of a Li-Ion battery under pulsed discharge, for which the recovery effect is modeled as a decreasing exponential function of the SOC and discharge capacity" [29].

1.2.5. Linear Equivalent Electric Circuits and Simscape Battery Models

Finally, the last category of models is that of the linear equivalent electric circuit models (ECM), such as those discussed in the next section, Much details on these category of the models can be found in [21] for a Li-ion polymer (LiPb). The last category of models are Simscape models, described also in the next section of the present research work. The weakness and the strengths of ECM and Simscape models concerning the SOC accuracy and robustness are discussed in detail in next section.

In conclusion, this article focuses on the design and implementation of two accurate SAFT Li-ion battery models, suitable for HEV applications. For each Li-ion battery model are implemented in Part 2 three real-time SOC estimators on a MATLAB R2020a platform. The remaining sections of this paper are structured as follows. Section 2 describes the first RC ECM model attached to a SAFT Li-ion battery. Section 3 describes the second Li-ion battery model, a Simscape nonlinear model. Section 4 analyzes the SOC performance through six statistical criteria. Section 5 details the authors' contributions to this research paper.

2. Li-Ion RC Battery Equivalent Circuit Model—Case Study and ADVISOR Setup

The purpose of this section is to present the case study of a small urban hypothetic car (SMCAR) which is set up using the ADVISOR 3.2 version software package, one of the most used in the automotive industry. Then, in next subsections is developed and validate an accurate Li-ion battery model that describes the dynamics of a SAFT Li-ion battery with a rated capacity of 6Ah and a nominal voltage of 3.6 V. This model is a third-order RC equivalent circuit model (3RC ECM), one of the most used in HEV applications due to its simplicity, high accuracy, and fast real-time implementation [14,17–23].

2.1. Li-Ion SAFT Battery and ADVISOR Small Hybrid Electric Car (SMCAR) Setup

SAFT is one of the most prestigious research companies in the US, among the most famous battery players on the commercial market in the world. It operates "under the auspices of the United States Advanced Battery Consortium (USABC) and the New Generation Vehicle Partnership (PNGV)," developing high-power lithium-ion (Li-ion) batteries over the past two decades. These batteries currently equip most HEVs and EVs [14–19]. The Li-ion battery together with other key components of a Hydrogen fuel cell electric vehicle are distributed on the car chassis as shown in Figure 1.

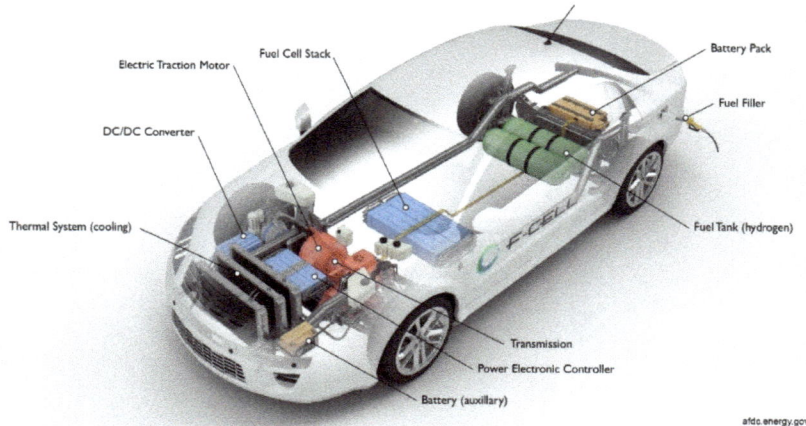

Figure 1. The distribution of the components on the car chassis (see [3], NREL).

The key components of a Hydrogen fuel cell electric car shown in Figure 1 are described in [2] as follows:

(1) The battery (auxiliary): this "powers vehicle accessories" and "provides electricity to start the car when traction battery is engaged".
(2) Battery pack: this "stores energy generated from regenerative braking and provides supplemental power to the electric traction motor".
(3) Direct current–direct current (DC–DC) converter: this is an electronic device that "converts higher-voltage DC power from the traction battery pack to the lower-voltage DC power needed to run vehicle accessories and recharge the auxiliary battery".
(4) Electric traction motor (FCEV): this is powered by the fuel cell and traction battery pack to drive the vehicle's wheels. It is also possible that "some of vehicles use motor generators that perform both the drive and regeneration functions".
(5) Fuel-cell stack: it is "an assembly of individual membrane electrodes that use hydrogen and oxygen to produce electricity".
(6) Fuel filter: this is "a nozzle from a high-pressure dispenser attaches to the receptacle on the vehicle to fill the tank".
(7) Fuel hydrogen tank-it "stores the hydrogen gas on board the vehicle until it is required by the fuel cell".
(8) Power electronics controller (FCEV): this is a unit that "manages the flow of electrical energy delivered by the fuel cell and the traction battery, controlling the speed of the electric traction motor and the torque it produces".
(9) Thermal system (cooling) (FCEV): this "maintains a proper operating temperature range of the fuel cell, electric motor, power electronics, and other components".
(10) Transmission (electric): this "transfers mechanical power from the electric traction motor to drive the wheels".

Among the Li-ion batteries of an HEV, the one with a capacity and a nominal voltage of 6 Ah and 3.6 V respectively is used for experimental validation tests, using an advanced simulator (ADVISOR) created in November 1994 by the US National Renewable Energy Laboratory (NREL). ADVISOR has so far proved to be the most suitable tools used in the design of HEV and EV systems, very well documented in [4–7]. Thanks to a wide variety of HEVs and EVs and the multitude of "real-world" driving conditions, it has gradually improved the performance until it reached version 2003-00, as well as the latest version r0116 of 24 April 2013, as mentioned in [14–18]. After proper installation, the ADVISOR graphical user interface (GUI) is running by typing "advisor" at the command prompt in MATLAB [14–18]. The ADVISOR GUI file menu has "help buttons which will either access the MATLAB help window or open a web page with appropriate context information" [15,16]. By using the ADVISOR GUI software package for design the following steps are requested:

Step 1. Define a vehicle.

Step 1.1. Define the input HEV page setup shown in Figure 2, based on a large collection of HEVs types and characteristics contained by software.

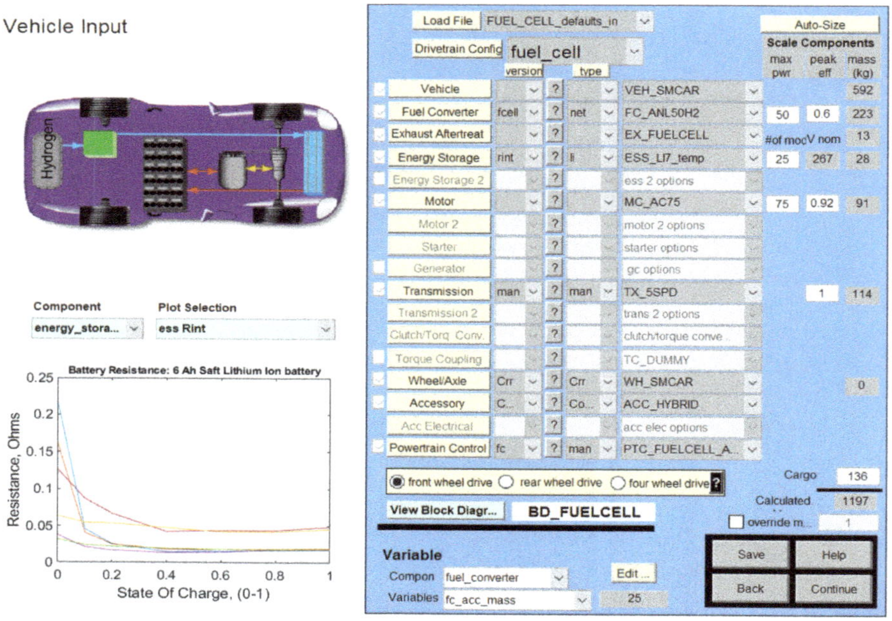

Figure 2. Vehicle input page setup.

As a case study we consider a hypothetical SMCAR, powertrain control hybrid (hydrogen fuel cell electric vehicle) with the following characteristics [16]:

- Fuel converter ANL Model 50 kW (net) ambient pressure hydrogen fuel-cell system.
- Transmission—5-speed manual transmission.
- Motor—Westinghouse 75 kW (continuous) alternating current (AC) induction motor/inverter.
- Accessory—type constant, i.e., 700 W constant electrical load, based on data from the SMCAR tests.
- Wheel/axle—wheel/axle assembly for small car SMCAR.
- Powertrain control—hybrid/adaptive control.
- Battery: rint (internal resistance)—type SAFT Li-ion battery 6 Ah-rated capacity and 3.6 V-nominal voltage. The characteristics of the SAFT battery are presented in Tables 1 and 2 [14,15,17–19]. Within a pack 84 cells are connected in series, with a nominal voltage capable of driving the PMDCM at approximately 300 V.

Table 1. Li-ion electrical characteristics for a SAFT battery of 6 Ah, 3.6 V [16].

Parameter/Coefficient	Symbol	Value	Unit Measure
Nominal Voltage	V_{nom}	3.6	V (volt)
Max charge voltage	V_{max}	3.9	V
Min discharge voltage	V_{min}	2.1	V
Nominal Capacity	C_{nom}	6	Ah (Ampere-hour)
Specify energy	η	64	Wh/kg
Energy density	η_{ch}	135	Wh/dm^3
Coulombic coefficient -for charging current -for discharging current	$\eta dsch$	0.98 0.86	

Table 2. Li-ion mechanical and thermal characteristics SAFT battery of 6 Ah, 3.6 V (cylindrical shape) [16].

Parameter/Coefficient	Symbol	Value	Unit Measure
1. Diameter	D	0.047	m (meter)
2. Height	h	0.104	m
3. Weight	G	0.375	kg
4. Volume	V	0.000018	m^3
5. Operating temperature range	t	(−10) (+45)	°C (degree Celsius)
6. Ambient temperature	t_0	20	°C
7. Heat capacity	cp	925	J/kg K ***
8. Heat transfer coefficient	hc	125	$W/(m^2 K)$
9. Constanta perfect gas	R	8.314	J/(mol K)
10. Thermal resistance	R_{th}	6	°C/W
11. Power losses	P_{loss}	1.9	W (watts)
12. Thermal time constant	T_c	2000	s
13. Activation energy	E	20,000	J/mol **

* J is the unit measure for mechanical energy (Joule), ** mol stands for number of molecules, *** K-stands for Kelvin degree.

The Simulink block diagram of the transmission system and Li-ion battery storage is shown in Figure 3.

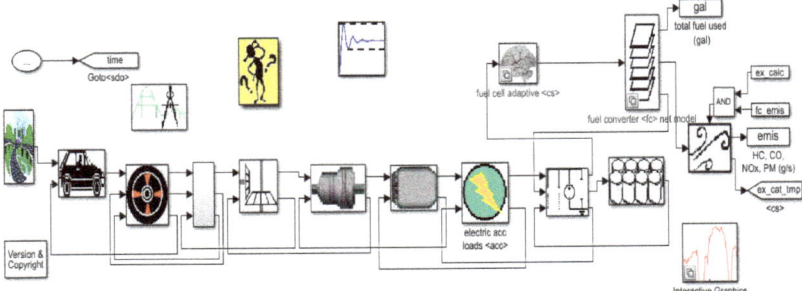

Figure 3. Simulink diagram of the small hybrid electric vehicle (HEV SMCAR) transmission system.

Step 1.2. Drivetrain selection—selects the drivetrain configuration of the vehicle (Series, Parallel, etc.).

Step 1.3. Selecting components.

Step 1.4. Editing variables.

Step 1.5. Loading and saving vehicle configuration.

Step 2. Running simulation.

Step 2.1. Select the drive cycle—in the case study we chose the Federal Test Procedure (FTP) driving cycle used by US Environmental Protection Agency (EPA) for emissions certifications of passengers' vehicles in USA. The FTP-75 shown in Figure 4 and converted in current profile charging and discharging cycle in Figure 5 is the standard federal exhaust emissions driving cycle, which uses an Urban Dynamometer Driving Schedule (UDDS [14–18]). The FTP cycle has three separate phases: one cold-start phase (505 s), followed by a hot transient phase (870 s) and a hot-start phase (505 s) [14,16,18]. For a 10 min cool-down period between second phase and the third phase the engine is turned off. The first and third phase are identical. The total test time length for the FTP is 2457 s (40.95 min). The top speed is 91.25 km/h and the average speed is 25.82 km/h. The distance driven is approx. 17.7 km [14,16].

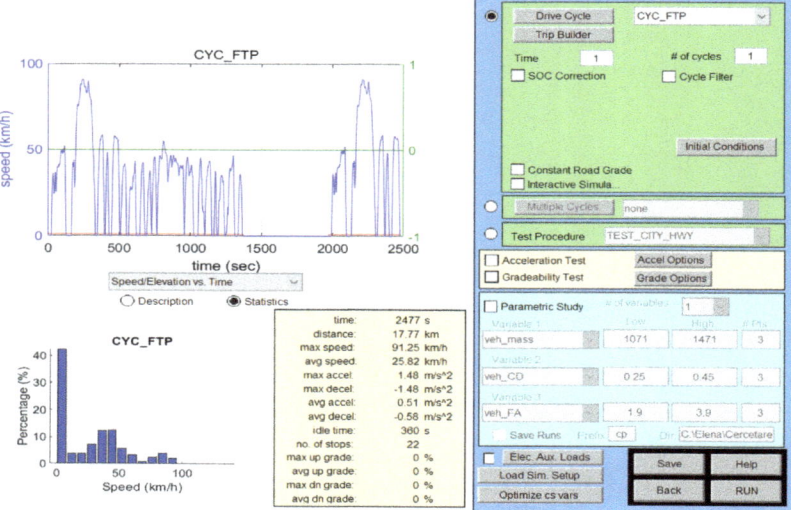

Figure 4. Simulink setup page.

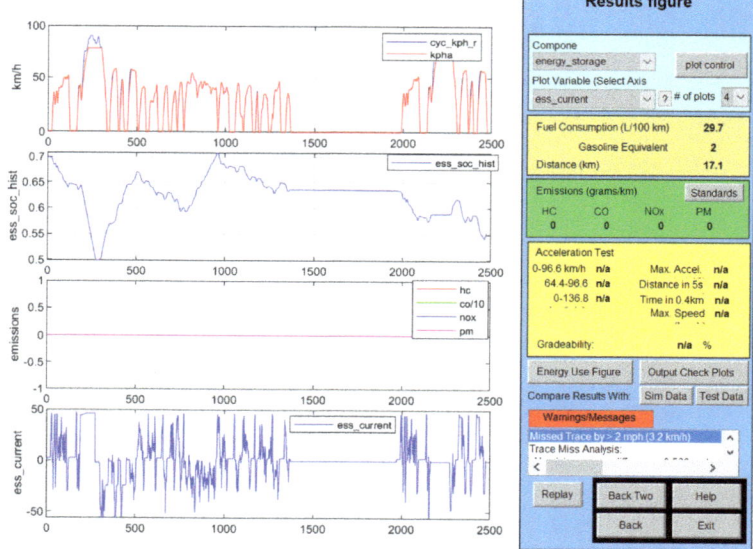

Figure 5. The Simulink simulation results.

Step 2.2. Select a trip builder for a repeated cycle (if the case).

Step 2.3. Select a SOC correct options (linear or zero delta).

Step 2.4. Select interactive simulation a real-time interactive simulation interface to activate while the simulation is running.

Step 2.5. Select multiple cycles to speed up the process of running many different cycles with the same initial conditions using this functionality.

Step 2.6. Choose a test procedure to select what kind of test to run.

Step 2.7. Save Simulink setup.

Step 2.8. Run the simulation and wait for the results figure to popup, as shown in Figures 4 and 5.
Step 3. Looking for the simulation results.

The first graph at the top of the Simulink configuration page, shown in Figures 4 and 5 (the first graph at the top), shows the ADVISOR FTP-75 driving cycle speed profile as the input variable. In Figure 5, the second graph at the top shows the estimated ADVISOR SOC value required to validate both models of Li-ion batteries attached to the SAFT Li-ion battery. The last chart at the bottom of Figure 5 shows the conversion of the FTP-75 driving cycle speed profile to a driving cycle current profile, required in MATLAB simulations for model validation and SOC estimators, as an input variable.

The SAFT Li-ion battery electrical characteristics specifications are given in Table 1, and Table 2.

2.2. Li-Ion-RC Equivalent Circuit Model

The 3RC ECM Li-ion battery model, as shown in Figure 6, consists of an OCV source in series with the internal Rin resistance of the battery and three parallel RC bias cells. RC cells are introduced into the circuit to capture the dynamic electrochemical behaviour of the battery and to increase the accuracy of the model. The first RC polarization cell captures the fast transient of the battery cell, and the last two RC cells capture the slow variation of the steady-state and increase the accuracy of the model. As the new technologies are largely dependent on batteries, it is important to develop accurate battery models that can be conveniently used with on-board power simulators and electronic on-board power systems, as mentioned in [14,17–23].

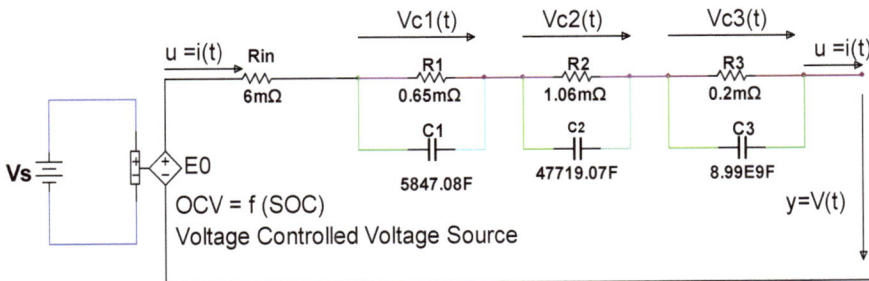

Figure 6. The third order 3RC electric circuit model (ECM)–Li-Ion battery represented in NI Multisim 14.1 editor (see [19]).

For simulation purpose, a specific setup for the 3RC ECM Li-ion battery model parameters, shown in Table 1 or directly on the electrical scheme from Figure 1, is available to prove the effectiveness of the proposed SOC estimation strategies. This setup is achieved from a generic ECM by changing only the values of the model parameters in state-space equations.

The Li-ion battery 3RC ECM model parameters are given in Table 3, and the OCV nonlinear model coefficients are shown in Table 4.

Table 3. The RC ECM parameters [17–23].

Item Measure	Parameters/Coefficients	Symbol	Value	Unit
	Li-ion battery ECM parameters			
1	Internal ohmic resistance	R_{in}	6	$m\Omega$ (milliohm)
2	First cell polarization resistance	R_1	0.65	$m\Omega$
3	Second cell polarization resistance	R_2	1.06	$m\Omega$
4	Third cell polarization resistance	R_3	0.2	$m\Omega$
5	First cell polarization capacitance	C_1	5847.08	F (Farad)
6	Second cell polarization capacitance	C_2	47,719.07	F
7	Third cell polarization capacitance	C_3	8.99×10^9	F

Table 4. RC ECM Li-ion battery open-circuit controlled voltage (OCV) coefficients.

Parameter/Coefficient	Symbol	Value
	K_0	4.23
	K_1	0.000038
	K_2	0.24
OCV coefficients	K_3	0.22
	K_4	-0.04

2.2.1. RC ECM Li-Ion Battery Model-Continuous Time State Space Representation

In a state-space representation, the continuous time 3RC ECM nonlinear model of SAFT Li-ion battery shown in Figure 6 is given by following Equations:

$$\frac{dx_1}{dt} = \frac{1}{R_1 C_1} x_2 + \frac{1}{C_1} u(t), \; u(\tau) \geq 0 \quad (2)$$

$$\frac{dx_2}{dt} = \frac{1}{R_2 C_2} x_2 + \frac{1}{C_2} u(t) \quad (3)$$

$$\frac{dx_3}{dt} = \frac{1}{R_3 C_3} x_3 + \frac{1}{C_3} u(t) \quad (4)$$

$$\frac{dx_4}{dt} = -\frac{\eta u(t)}{C_{nom}}, \; u(\tau) \geq 0 \quad (5)$$

$$OCV(t) = K_0 - K_2 x_4 - \frac{K_1}{x_4} + K_3 \ln(x_4) + K_4 \ln(1 - x_4) \quad (6)$$

$$y(t) = OCV(t) - x_1 - x_2 - x_3 - R_{in} u(t) \quad (7)$$

where the components of the state vector are: $x_4 = SOC$ is the state of charge of Li-ion battery, $x_1 = V_1$ is the voltage across first $R_1 \| C_1$ polarization cell, $x_2 = V_2$ denotes the voltage across the second $R_2 \| C_2$ polarization cell, $x_3 = V_3$ represents the third $R_3 \| C_3$ polarization cell $u(t) = i(t)$ is the input discharging current ($u(t) \geq 0$) or charging current ($u(t) \leq 0$), $OCV(t)$ represents the open-circuit voltage of Li-ion battery, and finally $y(t)$ designates the terminal voltage of the battery. The open-circuit voltage of Li-ion battery $OCV(t)$ given in (6) is a non-linear function of battery SOC, and contains a combination of the following three well-known generic battery models [17,19–21,25]:

(1) Shepherd model

$$y(t) = K_0 - R_{in} u(t) - \frac{K_1}{x_4} \quad (8)$$

(2) Unnewehr universal model

$$y(t) = K_0 - R_{in}Ru(t) - K_2x_4 \tag{9}$$

(3) Nernst model

$$y(t) = K_0 - R_{in}u(t) + K_3\ln(x_4) + K_4\ln(1-x_4) \tag{10}$$

The performance of the generic models in terms of voltage prediction and SOC estimation is analysed in [24], and the simulations result show that the Unnewehr and Nernst models compared to the Shepherd model, criticized in literature, increase significantly the accuracy of linear ECMs, more specifically, Nernst model "showed the best performance among the three mathematical models" due to its flexibility by using two parameters (correction factors instead of one). Last, the combination of all three mathematical models in (7) and their introduction in the terminal voltage relationship (8) increases considerable the Li-ion ECM accuracy. Also, the ECM combined model proved until now that it is "amongst the most accurate formulations seen in literature from EVs/HVs field" [17,21].

Since the parameters of 3 RC ECM Li-ion model strongly depend on temperature and SOC, the combined model is beneficial due to its simplicity, accuracy, and development of BMS SOC estimators for HEVs as a "proof concept" and fast real-time implementation.

It is important to underline that the values of coefficients K_0, K_1, K_2, K_3, and K_4, provided in Table 2, are chosen to fit the Li-ion battery model accurately according to the manufacturers' data by using a least squares curve fitting estimation method, as is suggested in [17–23]. The values of the resistances R_1, R_2 and the capacitors C_1, C_2, as well the value of the battery nominal capacity C_{nom} and its internal resistance are given in Table 1. The Simulink diagram of third order 3RC EMC–Li-Ion battery model that implements the Equations (3)–(8) is shown in Figure 7.

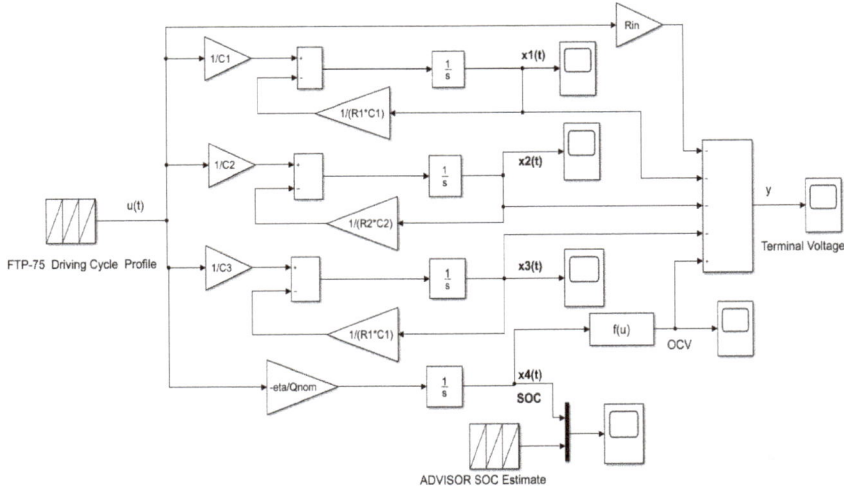

Figure 7. The Simulink diagram of third order 3RC ECM–Li-Ion battery model.

2.2.2. RC Electric Circuit Model (ECM)–Li-Ion Battery Model—Discrete Time State Space Representation

For the design and implementation of SOC estimators it is necessary to discretize over time the continuous model of the Li-ion battery. The discrete model of 3RC ECM Li-ion model is described in a compact state space matrix representation, as:

$$x(k+1) = Ax(k) + bu(k), \ u(k) \geq 0 \tag{11}$$

$$OCV(k) = K_0 - K_2 x_4(k) - \frac{K_1}{x_4(k)} + K_3 \ln(x_4(k)) + K_4 \ln(1 - x_4(k)) \tag{12}$$

$$y(k) = OCV(k) - x_1(k) - x_2(k) - x_3(k) - R_{in}u(k) \tag{13}$$

where $x(k) = \begin{bmatrix} x_1(k) \\ x_2(k) \\ x_3(k) \\ x_4(k) \end{bmatrix} \triangleq x(kT_s)$ is the state vector with 4 components $A = \begin{bmatrix} 1-\frac{T_s}{T_1} & 0 & 0 & 0 \\ 0 & 1-\frac{T_s}{T_2} & 0 & 0 \\ 0 & 0 & 1-\frac{T_s}{T_3} & 0 \\ 0 & 0 & 0 & 1 \end{bmatrix}$ denotes the state matrix, $b = \begin{bmatrix} \frac{T_s}{C_1} \\ \frac{T_s}{C_2} \\ \frac{T_s}{C_3} \\ -\frac{\eta}{C_{nom}} \end{bmatrix}$ is a vector of the input coefficients (input excitation terms), $u(k) \triangleq u(kT_s)$, $OCV(k) \triangleq OCV(kT_s)$, and $y(k) \triangleq y(kT_s)$ denotes the input, respectively the OCV and the terminal output voltage at discrete time instants $\{kT_s|_{k\in Z+}\}$. In our MATLAB simulations, the sampling time is set to $T_s = 1$ s, without any solver convergence problems.

2.2.3. Li-Ion Battery Thermal Model

Following the development in [14] the dynamics of Li-ion battery thermal model block is described by the following equations:

$$mc_p \frac{dT_c}{dt} = hS(T_0 - T_c) + R_{in}u^2 \tag{14}$$

$$T_c(s) = \frac{R_{th}P_{loss} + T_0}{T_{th}s + 1}, \quad P_{loss} = R_{in}u^2 \tag{15}$$

$$R_{in}(T) = R_{in}|T_0 + \exp\left(\propto \left(\frac{1}{T_c} - \frac{1}{T_0}\right)\right), \propto = \frac{E}{RT} \tag{16}$$

$$K_p(T) = K_p|T_0 + \exp\left(\beta\left(\frac{1}{T_c} - \frac{1}{T_0}\right)\right), \beta = \frac{E}{RT} \tag{17}$$

where, the variables and the coefficients have the following significance and values:

m the mass of the battery cell [kg]
c_p the specific heat capacity [J/molK]
S—the surface area for heat exchange [m^2]
T_c the variable temperature of the battery cell [K]
T_0 the ambient or reference temperature [K]
$R_{in}(T)$ the value of internal resistance of the battery cell dependent on temperature [Ω]
$K_p(T)$ ppolarization constant of Li-ion Battery [V]
u the input charging and discharging profile current [A]
$T_c(s)$ the internal temperature of the cell [°K] in complex s–domain (the Laplace transform)
R_{th} thermal resistance, cell to ambient (°C/W), $R_{th} = 6$ [°C].
T_{th} the thermal time constant, $T_{th} = 2000$ [s]
P_{loss} the overall heat generated (W) during the charge or discharge process [w]
$\propto, \beta$ Arrhenius rate constant
E—the activation energy, $E = 20$ [kJ/mol]
R Boltzmann constant, $R = 8.314$ [J/molK]

In MATLAB simulations, the battery temperature profile, and the robustness of the proposed SOC battery estimators are tested for the following approximate values, closed to a commercial battery type ICP 18,650 series [14]:

$$S = 15.4E - 3 \left[m^2\right], m = 0.0375 \text{ [kg]}, c_p = 925 \text{ [J/kgK]}, h = 5[w], t_0 = 20 \text{ [°C]} \tag{18}$$

An accurate simplified thermal model is provided in MATLAB R2019b library, at MATLAB/Simulink/Simscape/Battery, for a Li-ion generic battery model, implemented in Simulink Simscape as is shown in Figure 8, as is developed in [14]. The Equations (17) and (18) reveal a strong dependence of the internal resistance $R_{in}(T)$ and the polarization constant $K_p(T)$ of the Li-ion battery Simscape model.

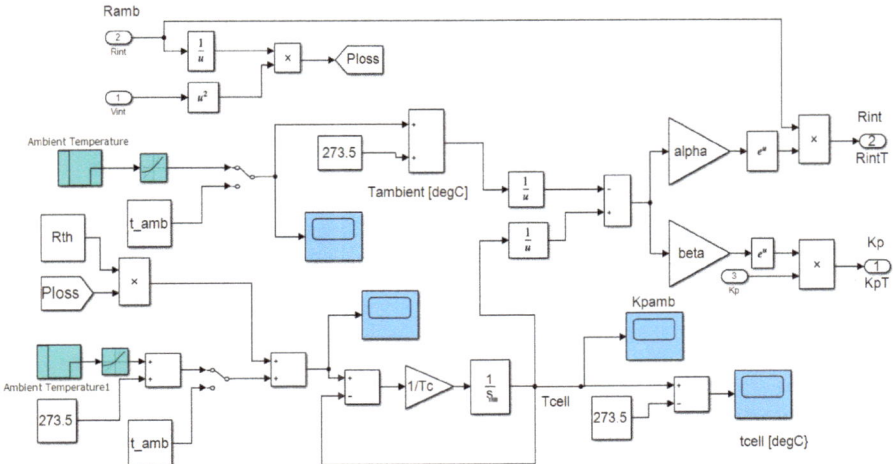

Figure 8. The detailed Simulink diagram of the Simulink Simscape thermal model block (see [14]).

Since the internal resistance of the Li-ion battery is the most sensitive to temperature developed inside the Li-ion battery, an overall Simulink model diagram block is designed that also integrates the Li-ion battery models such as is shown in Figure 9. It is essential to emphasize the fact that for performance comparison purpose, the overall Simulink model diagram shown in Figure 9, is sharing the same simplified thermal model to have an identical profile temperature and values of internal resistance $R_{in}(T)$ and polarization $K_p(T)$.

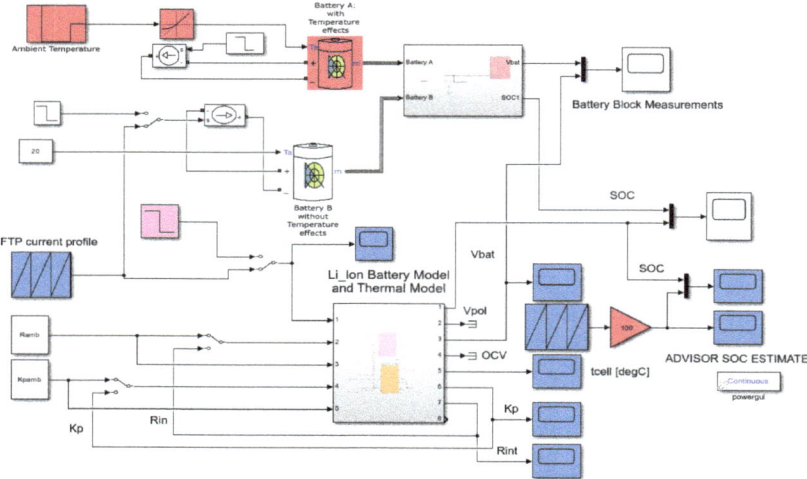

Figure 9. Simulink Simscape model diagram setup that integrates two main blocks. Legend: First block from bottom side encapsulates the Li-ion battery model and Simulink thermal model block; Second block from the top side is a Simscape block with two Li-ion batteries, first one from the top simulate the temperature effects and second one from the bottom of first one doesn't take into consideration the temperature effects.

It is important to remark that in Figure 9 the second block from the top of the Simulink diagram is introduced only to investigate the SOC and temperature profile evolutions delivered by the first block from the bottom side in comparison to SOC and the temperature profile delivered by the top side block. The ambient temperature profile and the output temperature of the Simulink Simscape thermal model described by Equations (15)–(18) are shown in Figure 10a,b respectively.

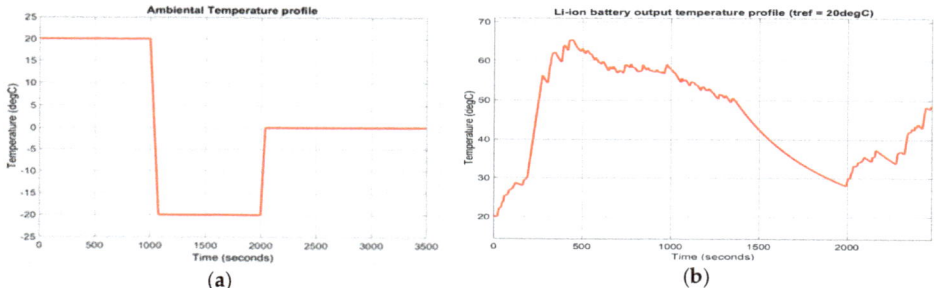

Figure 10. (**a**) The input ambient temperature profile; (**b**) the output temperature as response to input ambient temperature of the thermal model block.

The evolution of the battery internal resistance $R_{in}(T)$ and polarization constant $K_p(T)$ at room temperature $t_0 = 20\ °C$, is shown in Figure 11a,b.

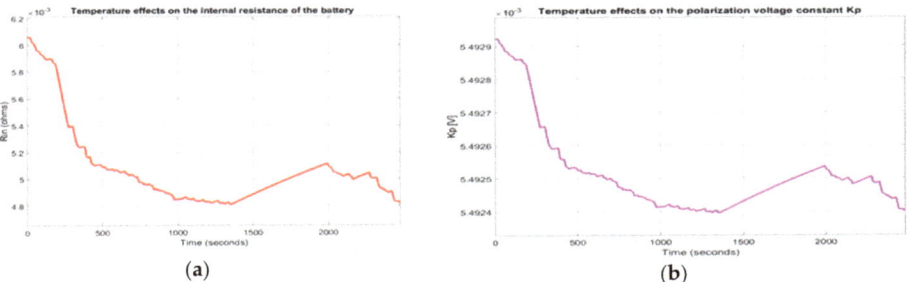

Figure 11. (a) The internal battery $R_{in}(T)$ at ambient temperature (20 °C); (b) The polarization constant at ambient temperature $K_p(T)$ (20 °C).

The output temperature profile of the Simulink Simscape thermal model for changes in ambient temperature is shown in Figure 12a, and the effects on internal battery resistance $R_{in}(T)$ and polarization $K_p(T)$ are presented in Figure 12b,c.

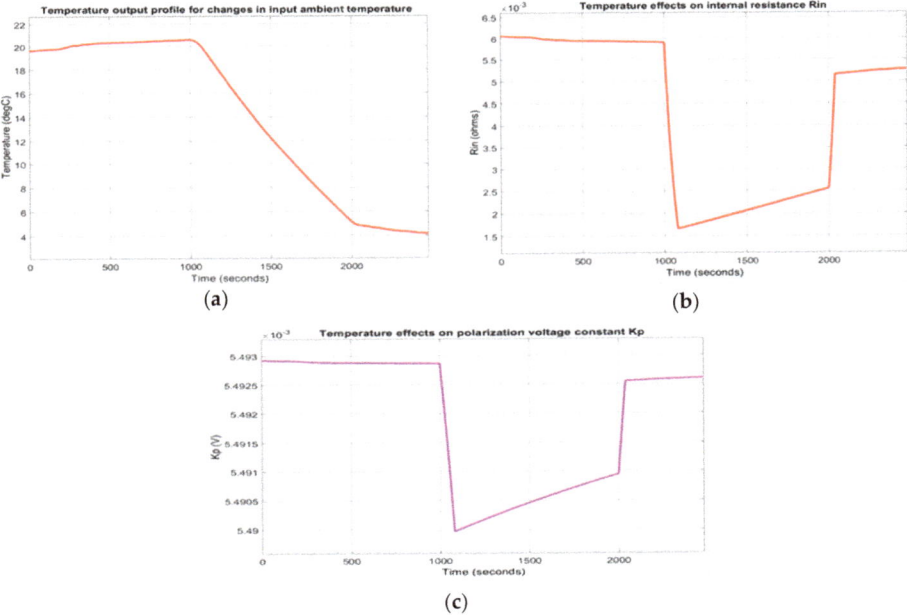

Figure 12. (a) The effect of changes in ambient temperature, shown in Figure 10a, on output temperature profile; (b) The effect of changes in ambient temperature on the internal resistance $R_{in}(T)$; (c) The effect of changes in ambient temperature on the polarization constant $K_p(T)$.

2.2.4. RC ECM Li-ion Battery Model—MATLAB Simulink Simulations Result

The MATLAB simulations result of 3RC ECM Li-ion battery model implementation is shown in the Figures 13 and 14. In Figure 13a,b are depicted the FTP current profile test (a), and the value of 3RC ECM Li-ion model SOC versus ADVISOR SOC estimate to the FTP current profile test (b) obtained on NREL ADVISOR MATLAB platform as shown in Figure 5.

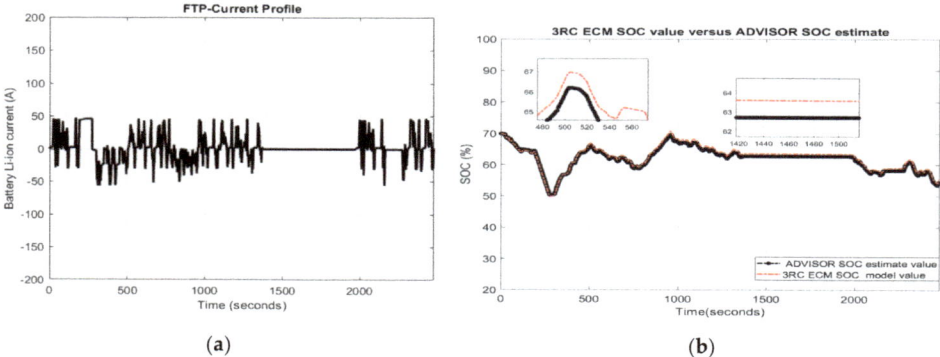

Figure 13. 3RC ECM Li-ion battery model (**a**) The Federal Test Procedure (FTP)-75 current profile test; (**b**) The corresponding 3RC ECM SOC true value versus ADVISOR state of charge (SOC) estimate (Li-ion battery model validation).

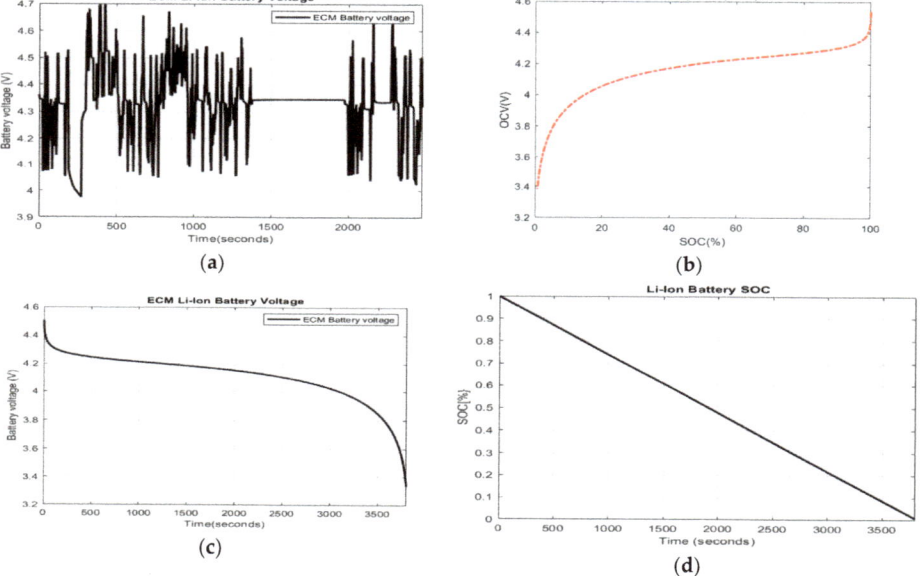

Figure 14. (**a**) Li-ion battery terminal voltage; (**b**) the OCV = f(SOC) curve for a discharging constant current at 1C-rate (6A); (**c**) the Li-ion battery voltage; (**d**) the Li-ion battery SOC for a discharging constant current at 1C-rate (6A).

Using a MATLAB magnification tool on two portions of the graph, the visibility of both curves shown in Figure 13b increases considerably [26]. The Li-ion battery terminal voltage for an FTP-75 current charging and discharging profile test, the following three, namely OCV = f(SOC) curve, battery terminal voltage and its SOC, all of these three simulated for a constant discharge current at 1C-rate (6A), it can see in the Figure 14a–d.

The simulation results from last three Figure 14b–d reveal that all three battery characteristics are quite close to the manufacturing specifications. It should also be noted that the OCV = f(SOC) curve from Figure 14b is almost flat on a large portion. Therefore, the Coulomb counting method is not accurate for direct SOC measurement for Li-ion batteries. Thus, its estimation is necessary using one of the best known Kalman filtration techniques.

2.2.5. RC ECM Li-Ion Battery Model—MATLAB Model Validation

For validation of 3RC ECM Li-ion battery model, in the first stage is calculated the model SOC residue as a difference between the SOC values of the 3RC ECM model and the estimated values of ADVISOR SOC estimator. The SOC accuracy performance of 3RC ECM Li-ion battery model is analyzed by evaluating the SOC residual error. The residual percentage error is depictured in Figure 15.

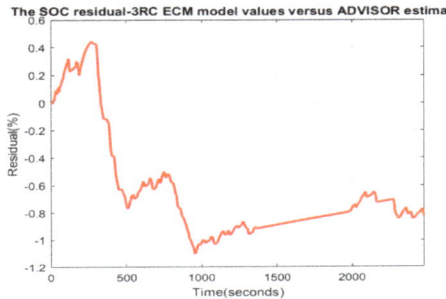

Figure 15. The Li-ion battery SOC accuracy performance assessment–SOC residual.

The results of the MATLAB simulations shown in Figure 13b for a current FTP-75 driving cycle profile test reveal excellent SOC accuracy of the 3RC ECM compared to the estimated SOC value obtained by the ADVISOR simulator. From a quantitative point of view, this confirms the information extracted by evaluating the SOC residues generated in MATLAB and presented in Figure 15. From Figure 15 it can be seen that the SOC residue is in the range [−1.1, 0, 4], and the SOC error rate is less than 1.2%, which is an excellent result, comparable to those reported in the literature, even better. This result reveals that the 3RC ECM Li-ion battery model is very accurate in terms of SOC calculation, and the model is undoubtedly validated based on available information about its behaviour.

3. Li-Ion Battery Simscape Generic Model

A full representation of the generic battery model, dependent on the temperature and ageing effects, is developed by MathWorks team, as shown in the MATLAB R2019b/Simulink/Simscape/Power Systems/Extra Sources Library-Documentation.

3.1. Li-Ion Battery Simscape Generic Model—Description and Parameters' Specifications

Li-ion battery cell specifications for a Simscape model are shown in Figure 16a–c. The Li-ion battery Simscape model is more realistic and suitable to operate safely in different conditions. Also, this model is beneficial for an appropriate choice of battery chemistry and for different parameters specifications. The Simscape generic model developed by MathWorks team takes into consideration the thermal model of the battery (internal and environmental temperatures) and its ageing effects. The battery terminal voltage, current and SOC "can be visualized to monitor and control the battery SOH condition" [14].

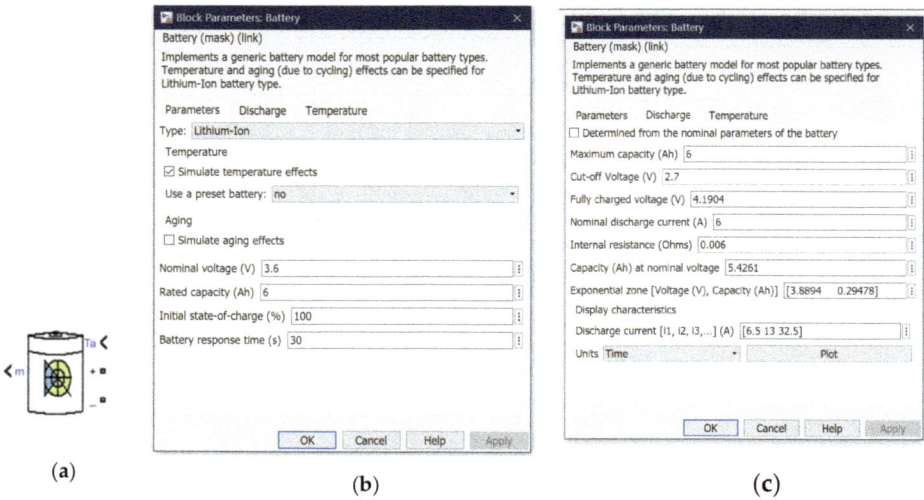

(a) (b) (c)

Figure 16. SAFT Li-ion battery specification—Simscape model; (**a**) Simscape model graphic representation (icon); (**b**) block parameters and battery type; (**c**) block parameters' battery specification for a discharging constant current.

The nominal current discharge characteristics according to a choice of a Li-ion battery which has a rated capacity of 6 Ah and a nominal voltage of 3.6 V for different x-scales (time, Ah) are shown in the Figures 17 and 18.

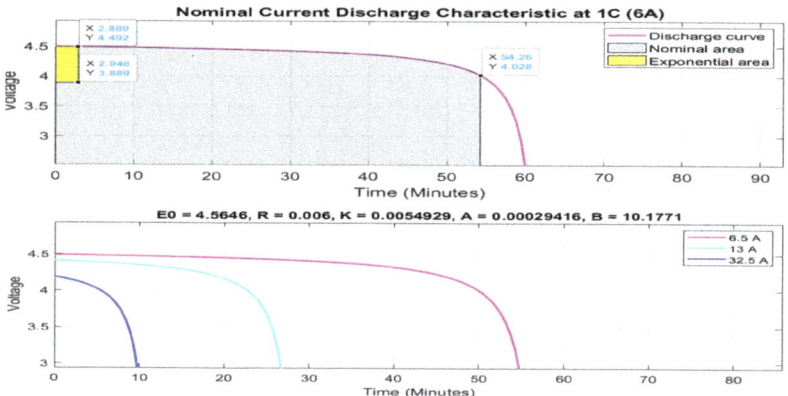

Figure 17. SAFT Li-ion battery nominal current discharge characteristic @1C (6A) (top side view); @6.5A, 13A and 32.5A (bottom view)—Simscape non-linear model (x-scale is the time in minutes).

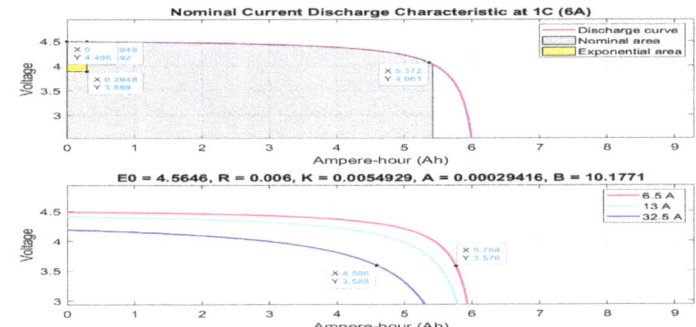

Figure 18. SAFT Li-ion battery nominal current discharge characteristic @1C (6A) (top side view); @6.5A, 13A and 32.5A (bottom view)—Simscape nonlinear model (x-scale is the capacity in Ampere-hour (Ah)).

The Simscape model of a generic 6 Ah and 3.6 V Li-ion battery SAFT-type without temperature and ageing effects is shown in Figure 19, the same shown in [14], p. 12.

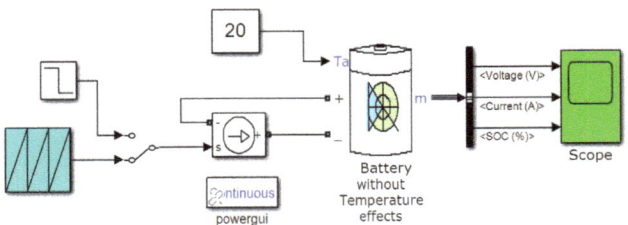

Figure 19. The Simscape model of a generic 6 Ah and 3.2 V Li-ion battery (without temperature and aging effects (see [14], p. 12) connected to FTP-75 input current profile.

A significant advantage of the Simscape model of Li-ion battery is the simplicity with which the model parameters for different chemistry and specifications are extracted as if we had access to the specifications of the battery manufacturers. The parameters of Li-ion battery choice extracted from the discharge characteristics shown in Figure 17 or Figure 18 have the following values:

$$E_0 = 4.5646 \ [V], R_{in} = 0.006 \ [\Omega], K_p = 0.0054929 \ [V], A = 0.00029416, B = 10.1771 \quad (19)$$

where

E_0 denotes the battery constant voltage [V].
R_{in} designates the internal resistance of the battery [Ω].
K_p is the polarization battery voltage constant [V].
A represents the exponential zone amplitude [V].
B means the exponential zone time constant inverse [1/(Ah)].

3.2. Li-Ion Battery Simscape Model—Discrete Time in State Space Representation

The Simscape model parameters suggested in Figures 16–18 fit the following adopted model represented in discrete time in a unidimensional state space, like the model developed in [14], p. 21:

$$x_1(k+1) = x_1(k) - T_s \left(\frac{\eta}{Q_{nom}}\right) \times u(k) \quad (20)$$

$$y(k) = E_0 - \frac{K_p T_s}{x_1(k)} \times u(k) + A\exp\left(-\frac{BQ_{nom}}{\eta}(1 - x_1(k))\right) - R_{in}u(k) \tag{21}$$

where $x_1(k) \triangleq x_1(kT_s) = SOC(kT_s)$, $u(k)$, $y(k)$, Q_{nom}, η and T_s have the same meaning as the variables and parameters that describe the 3RC ECM Li-ion battery model given by Equations (12)–(14). It is essential to emphasize a great advantage of the adopted Simulink Simscape model, presented in (21) and (22), consisting of a considerable model simplification and dependence only on SOC. Also, the dynamics of this model is described by the first Equation (21) which is linear and the second Equation (22) is a highly nonlinear static representation.

The Simulink Simscape model of Li-ion battery that implements Equations (21) and (22) is shown in Figure 20.

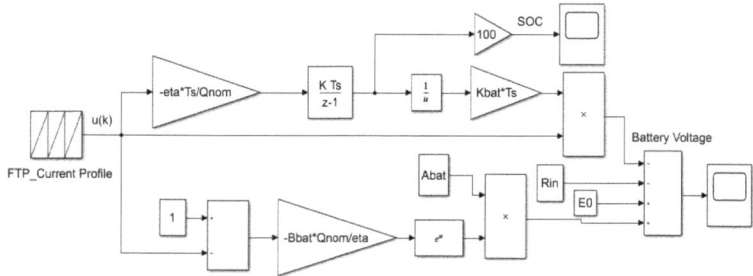

Figure 20. Simulink Simscape Diagram of Li-ion model. The values of the parameters from Simulink diagram are allocated in a MATLAB script that runs first for initialization, and then is running the Simulink model to extract these values from MATLAB workspace.

3.3. Li-Ion Battery Simscape Generic Model—MATLAB Simulations Results and Model Validation

The MATLAB simulations result is shown in Figure 21a,c.

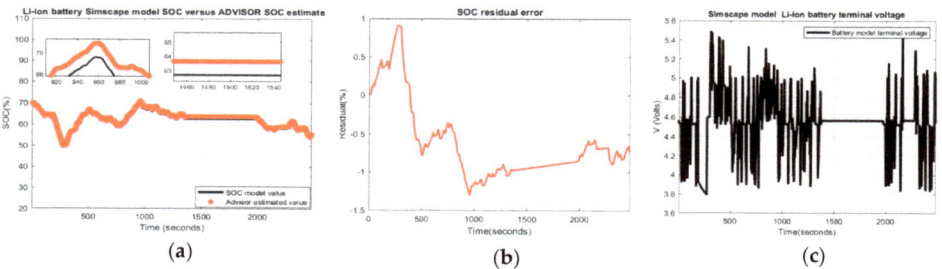

Figure 21. Simscape model Li-ion battery SOC accuracy assessment (**a**) Li-ion battery Simscape model SOC versus ADVISOR SOC estimate; (**b**) SOC residual; (**c**) terminal output voltage.

In Figure 21a, the simulation result reveals an excellent SOC accuracy of the Simscape model of the Li-ion battery. This result is also supported by a small SOC residue, recorded in Figure 21b, which falls in the range [−1.4, 1]. Like the 3RC ECM Li-ion battery model, the Simscape model of the Li-ion battery based on the available information extracted from Figure 21a,b also works very well, because the SOC error percentage is less than 1.4%, compared to the typical value of 2% reported in the literature for similar applications. These results also validate this model, which is suitable to use it in the second part for real-time design and implementation on an attractive MATLAB 2020Ra environment.

3.4. Simulink Simscape Graphic Models Integrated in Fuel Cell HEV Applications—Energy Management System

This section presents some HEV applications that operate with graphic Simscape models. In this description, the Simscape "blocks language" allows much faster models of physical systems to be created within the Simulink environment, "based on physical connections that directly integrate with block diagrams and other modeling paradigms" [8]. In Simscape, the models can be parametrized using MATLAB variables and expressions and can be designed and implemented control systems for any physical system in Simulink". Users can easily integrate physical object icons into the design of Simulink diagrams or combine object models with the symbols of different physical objects. Indeed, behind each image is encapsulated the dynamic pattern of physical objects. However, it significantly eliminates the user's effort to write a lot of equations for modelling the dynamics of objects, which takes a long time, and the diagrams become much more complicated [8].

3.4.1. Hybrid Energy Storage of Energy Management System (EMS)—Simulink and Simscape Components Description

The hybrid energy storage (HES) of an EMS, shown in Figure 22, is a hybrid combination of three power sources, such as a fuel cell, Li-ion battery, and supercapacitor [5,7]. The control strategy of HES is implemented in Simulink Simscape to "manage the energy consumption of the hydrogen fuel, and at the same time the pulsed or transient power required (load profile) by the load should be supplied." [11]. To simplify the Simulink diagram of the EMS, are used Simscape components such as Li-ion battery, supercapacitor and FCPM that also encapsulates a hydrogen fuel stack cell Simscape model, provided by MATLAB Simulink Toolbox/Simscape. In this section, a brief presentation of this topic is given, since is only emphasized the fact that using a single power source such as Li-ion battery in driving HEV powertrains applications "has certain disadvantages such as recharging, longevity, poor power density, etc." [24].

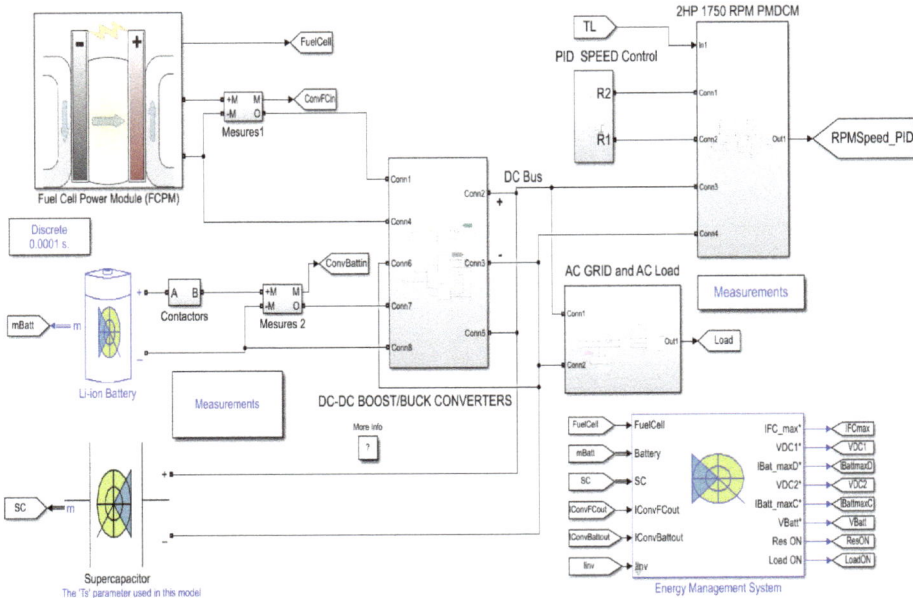

Figure 22. The adapted EMS of HEV SMCAR—Simulink Simscape diagram (adapted from Noya, [5,7]).

In this diagram other Simscape components are integrated such as three DC–DC boost/buck converters blocks to interface with all three sources. The first is a 12.5 kW fuel cell DC/DC boost converter, with regulated output voltage and input current limitation, and the other are two DC/DC converters for discharging (4 kW boost converter) and respectively for charging (1.2 kW buck converter) the battery system. Normally, a "single bidirectional DC/DC converter can also be used to reduce the weight of the power system" [7]. The FC is controlled by a DC–DC boost converter, an electronic device controlled also by a signal sent by one of five control strategies conceived for this purpose inside the EMS block, as is described in [5]. Similarly, the Li-ion battery and the supercapacitor are controlled by a bidirectional DC-DC buck-boost converters, since during operation they are charging and discharging. The topology configuration and the electronic circuits are well described in [11]. The charging and the discharging cycles of the bidirectional DC-DC converters are controlled by a voltage signal provided by EMS block that adjusts the duty cycles (D) of both DC–DC converters, based on the following relationship [11]:

$$D = \frac{V_{out}}{V_{out} - V_{in}} \quad (22)$$

where D is the duty cycle, V_{out} designates the output voltage of the converter, and V_{in} denotes the input voltage. In this section is presented only briefly the most relevant MATLAB simulation results for EMS techniques to have a better insight of the behavior of all three Simscape components of hybrid power sources, i.e., FC, Li-ion battery, and SC (UC). In the Simulink Simscape diagram of fuel-cell hybrid power generation (FCHPG) shown in Figure 22, the inverter DC/AC that supplies the load is rated at 270 V DC in input, and 200 V AC, 400 Hz, 15 kVA in output.

A three-phase load profile is "emulated to consider variations in power at the different timings and simulations to see the behavior of the hybrid energy storage system (HESS) as a whole and the response of each storage system" [11]. Also, a Simscape "15 kW protecting resistor is integrated in the Simulink diagram to avoid overcharging the supercapacitor and battery systems" [7].

3.4.2. Hybrid Energy Storage of EMS—Simulink Simscape Applications

As a practical application, the following three scenarios are implemented to reveal the behavior of HESS components:

- Scenario 1: DC grid interfaces only the AC Grid and AC load, such in [7,11].

The HESS distributes the power among the energy sources according to a given energy management strategy. The MATLAB simulation results of EMS techniques are shown only for three setups, such as the state machine control strategy (SMCS), classical PI control strategy (PICS), and the equivalent consumption minimization strategy (ECMS) [5,7,25]. To obtain a sound theoretical background on the EMS design and implementation in a real-time MATLAB simulation environment, the following sources [5,7,11,24] provide valuable information. For EMS-SMCS setup shown in Figure 23, the MATLAB simulations result is presented in the Figure 24.

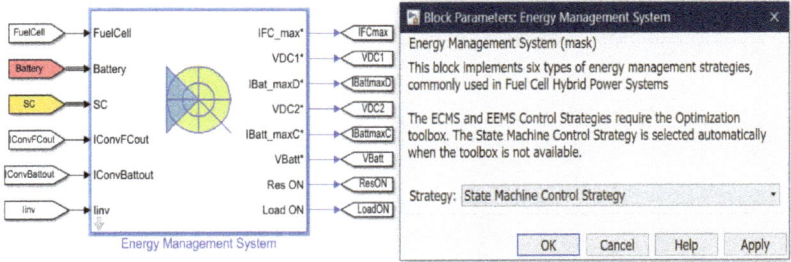

Figure 23. The EMS—state machine control strategy setup.

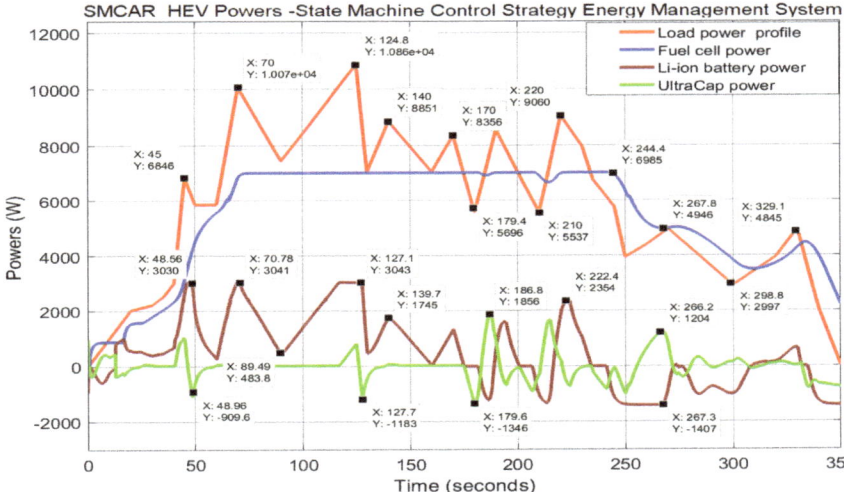

Figure 24. SMCAR HEV power–EMS state machine control strategy setup.

In Figure 24 is depictured the powers' distribution for FC, Li-ion battery, UC, and Load profile for a hypothetical SMCAR HEV case study. The balance equation is given by:

$$P_{FC} + P_{Batt} + P_{UC} = P_L, \tag{23}$$

where P_{FC} is the power provided by FC, P_{Batt} is the power delivered by Li-ion battery, P_{UC} is the power delivered by the UC to manage power peaks for vehicle acceleration and regeneration, and P_L is the load profile (demand, total power required). From Figure 24 it is straightforward to check that Equation (23) is satisfied for each time moment. Also, it is obvious that for load power profile pecks the power delivered by UC is very sharp to cover the power demanded (P_L).

In the Figure 25a–d are shown the load profile (a), fuel cell voltage (b), fuel current (c) and hydrogen consumption respectively (d), according to the load profile.

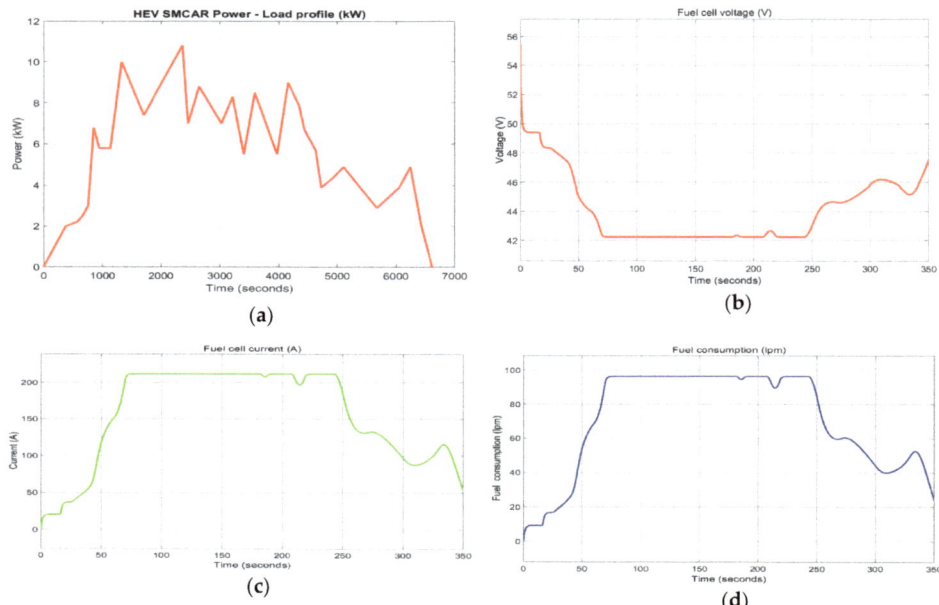

Figure 25. (a) Load profile; (b) fuel-cell voltage; (c) fuel-cell current; (d) fuel hydrogen consumption.

The simulation results in Figure 25 show a decrease in FC battery voltage from 56 V at t = 0 [s] to 42 V at t = 70 [s] and remain almost constant until t = 250 [s], followed by an increase to 48 V at t = 350 [s]. The FC current has the same evolution trend as the FC power, increasing from 0 [A], to t = 0, at approximately 210 [A] at t = 70 [s]. Inside the window [70, 250] [s] the FC current remains almost constant, and at t = 250 [s] it decreases to 80 [A] at t = 350 [s]. FC fuel consumption follows the same trend as FC current. The amount of fuel increases at the beginning of the simulation to 100 [lpm], then remains almost constant inside the window [70, 250] [s] when it delivers maximum power to the DC network, because in this interval the load power profile reaches some peaks of maximum value between 8 kW and 10 kW.

In the Figure 26a,b are depicted the UC current (a) and UC voltage variation (b) respectively, according to the load profile.

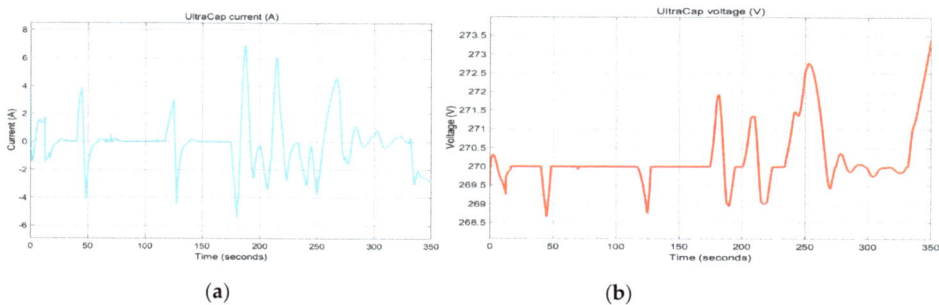

Figure 26. (a) UC current; (b) UC voltage.

Matlab simulation results reveal an evolution with sharp peaks for UC current and UC voltage when Li-ion battery needs to provide much more power to the DC network or during sudden acceleration and regeneration.

In the Figure 27a–c are presented the Li-ion battery current (a), battery terminal voltage (b) and battery SOC (c) respectively, according to power required.

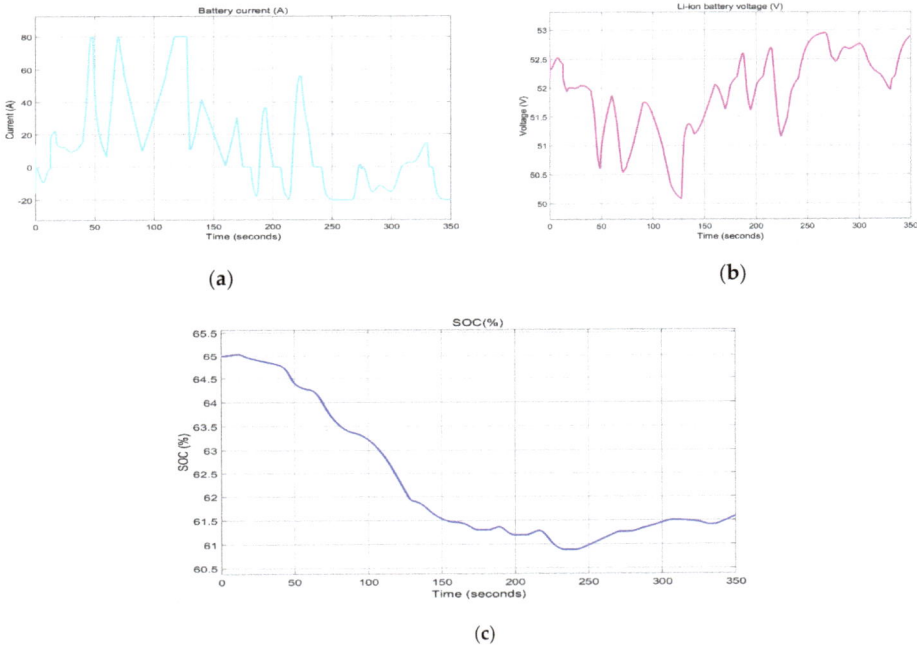

Figure 27. (a) Li-ion battery current; (b) battery terminal voltage; (c) battery SOC.

Figure 27a,b show several peaks (positive and negative) in the evolution of the current of the Li-ion battery, which correspond to the charging and discharging cycles of the battery, as can be seen from the evolution of SOC in Figure 27c. The terminal voltage of the Li-ion battery also has a lot of variations in its growth, decreasing from 83 V to t = 0 V to 60 V around t = 125 [s], followed by an increase to 83V when t = 83V.MATLAB simulation results analysis for Scenario 1

In this application, it is important to analyze the power distribution shown in Figure 24. The result of the analysis provides a better perspective on how EMS works in real-time simulations. In this figure, the red colour curve represents the profile of the power load, i.e., a variable power required for the AC load in the first 350 s of real-time simulation. The blue colour curve designates the main power generated by FC source, which is the dominant source, i.e., the one that delivers the most considerable amount of power to a DC grid and is almost constant inside the 150 s window length (75,225) [s]. The brown curve refers to the second power supply source, which is a Li-ion battery that delivers power to the DC grid in a smaller and variable amount during charging and discharging cycles, compared to FC. Finally, the green colour curve refers to the third power supply source that delivers the smallest amount of power to the DC grid only during the short periods of sharp acceleration and regeneration. The power distribution balance can be easily checked for enough moments because the MATLAB Data

Tips measurement tool can help to mark several points on each curve. For example, at time t = 70 s, the power delivered by each source of power supply has the following values:

$$P_{FC} = 6.985 \text{ kW}, P_{Batt} = 3.041 \text{ kW}, P_{UC} = 0 \text{ kW, and } P_L = 10.07 \text{kW}$$

The power distribution evaluated at t = 70 s verifies with enough accuracy Equation (23), since:

$$P_{FC} + P_{Batt} + P_{UC} = 10.026 \text{ kW, so close to } P_L = 10.07 \text{ kW}$$

As Equation (23) is satisfied for each moment, it is easy to observe the behavior of all three power supply sources. The MATLAB simulation results shown in Figures 25c, 26a and 27a reveal the same trend of current evolution as that of each corresponding power supply.

As in the case of the EMS-SMCS setup, similar graphs with the same meaning are presented in Appendix A, Figures A1–A10 for second EMS-PICS setup, and in the ([30], Figures A11–A20) for third EMS-ECMS setup. The theory and all the Simulink diagrams behind the five EMS techniques are fully documented in [5]. In our research, these EMS techniques are presented only as complementary information for interested readers, such as to give a clue, motivation and to open new research directions for future HEV developments. Nonetheless, the topic is beyond the scope of this paper, which is focused only on the modeling aspects and Li-ion battery SOC estimation techniques.

- Scenario 2: A 100 HP, 1750 RPM asynchronous induction motor (squirrel cage) is connected to AC grid as is shown in Figure 28.

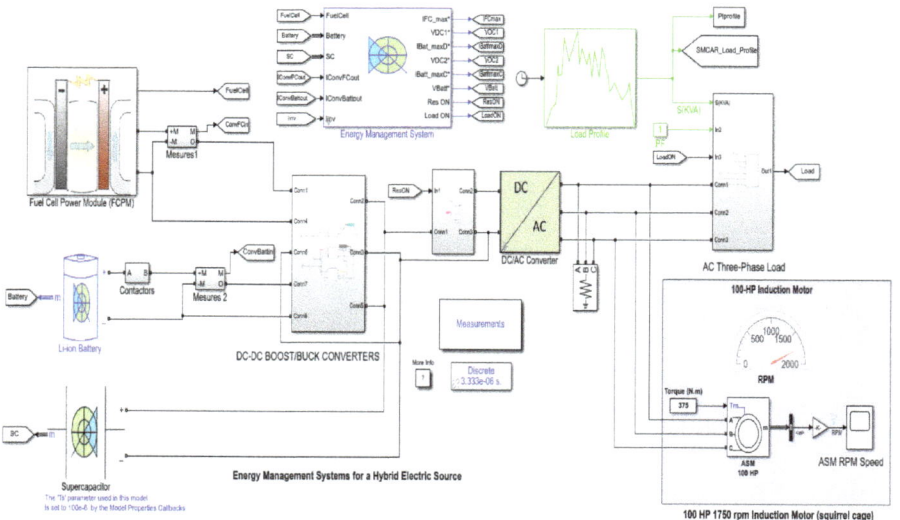

Figure 28. The 100 HP, 1750 RPM speed asynchronous induction motor (squirrel cage) connected to the output of a direct current/alternating current (DC/AC) converter (inverter).

In this scenario is shown only the MATLAB simulations result related to the evolution of asynchronous induction motor speed, as can be seen in Figure 29.

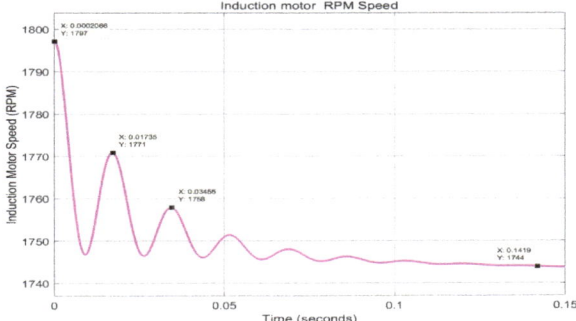

Figure 29. The MATLAB simulation result of unregulated RPM speed of induction motor.

From Figure 29, it is easy to observe that the speed of induction motor connected to AC grid is not controlled, but it is quite close to 1750 RPM in steady state, for a torque load of 375 Nm.

- Scenario 3: A 2 HP 1750 permanent magnet DC motor connected to DC side of the grid, as is shown in Figure 22 (the right topside block).

Since the block from Figure 22 encapsulates the PMDCM Simscape model, for clarity, Figure 30 shows the PMDCM Simscape model with all the details of electrical connections.

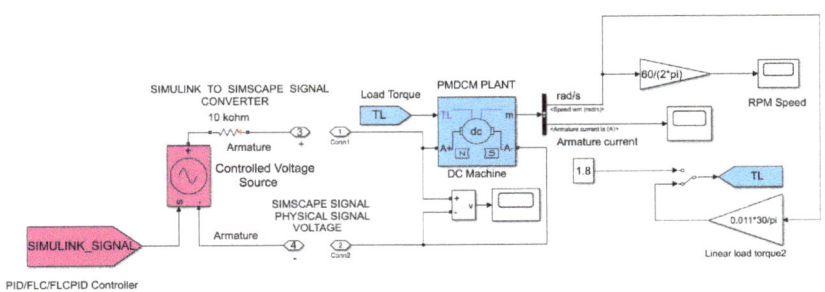

Figure 30. A 2HP 1750 RPM PMDCM—Simscape model (see [12]).

Unlike the uncontrolled speed of the asynchronous induction motor (ASM), the PMDCM is connected by a negative feedback in a closed-loop to a block of proportional-integral-derivative (PID) controller, to control its speed, as is shown in Figure 31, and developed in [12].

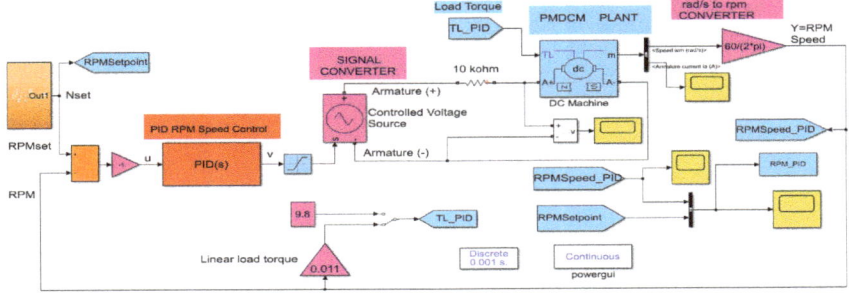

Figure 31. Closed-loop PMDCM-proportional-integral-derivative (PID) RPM speed control.

The MATLAB PMDCM RPM speed step response is shown in Figure 32. A big advantage of the PID controller is that the PMDCM speed response converges quickly and reaches the target speed of 1500 RPM in almost 1.8 s.

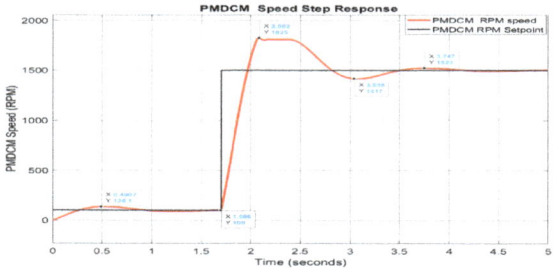

Figure 32. PMDCM RPM speed step response.

The characteristics curves of a Li-ion battery connected to PMDCM as a DC load are shown in Figure 33a–c, namely the battery SOC (a), a sequence of discharging and charging current cycles (b), and the battery terminal output voltage changes during battery operation (c). The SOC of the Li-ion battery remains almost constant during PMDCM operation. The battery current increases to almost 15 A at the beginning of the first transient and remains constant for a short period of time during steady state. However, at the beginning of the second transient it rises sharply to 60 A for a short time, then slowly decreases to −15 A at the end of steady state. The battery output voltage, shown in Figure 33c, decreases slightly during the first transient from 62.4 V to 62.2 V, and at the beginning of the second transient falls slowly for a short time and then increases linearly. At the end of steady state, it reaches 62.4 V. Therefore, the evolution of the battery output voltage is smooth, keeping an almost constant value during PMDCM operation.

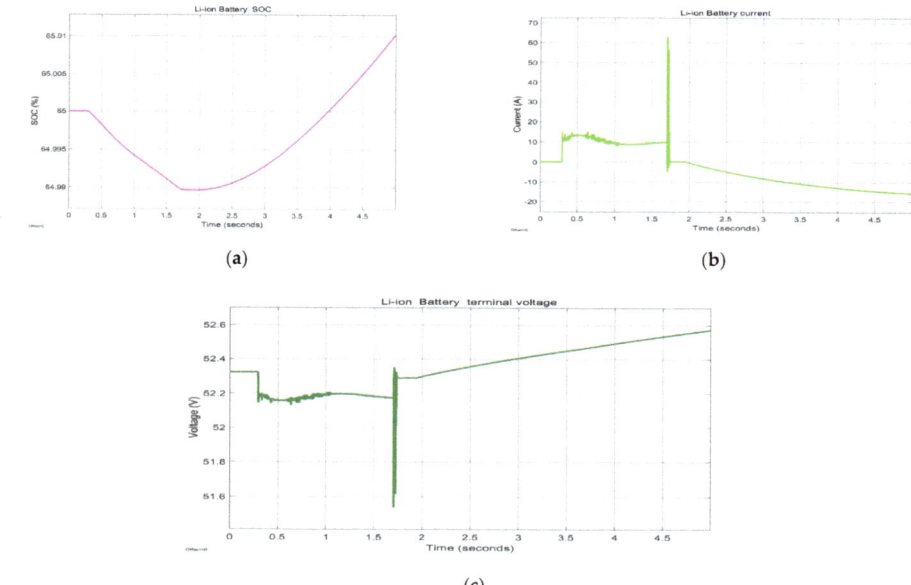

Figure 33. Li-ion battery supplying the PMDCM; (**a**) Li-ion battery SOC; (**b**) Li-ion battery current; (**c**) Li-ion battery terminal voltage.

The behaviour of the SC connected to the DC grid to supply power to the PMDCM during sudden changes in the load torque is described in Figure 34a for the SC current and in Figure 34b for the SC voltage. The SC current, during the first transient, decreases from 2000 A to 900 A at the end of the first steady-state period, and then at the beginning of the second transient continues the fall to −1000 A for which t = 2 [s]. After that the SC current increases to almost −300 A at t = 5 [s] which coincides with the end of steady-state. The SC voltage increases during the first transient from 0 [V] to almost 10 [V] at t = 1.8 [s], when PMDCM suddenly changes speed from 100 RPM to 1500 RPM, absorbing much more power. This justifies the presence of the SC to provide much more energy at this switching time. Therefore, the SC voltage suddenly increases from 10 V to 350 V in the time window (2, 2.5) [s] and then decreases to 200 V until the end of the steady-state. During this time, the SC protects the Li-ion battery so that it works smoothly, maintaining a constant SOC value.

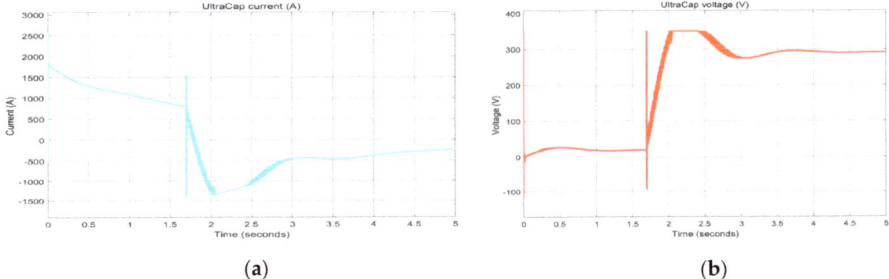

Figure 34. Supercapacitor (SC) supplying the PMDCM during sharp changes in the load torque; (**a**) SC current; (**b**) SC voltage.

The PMDCM behavior during the operation is shown in Figure 35a–d.

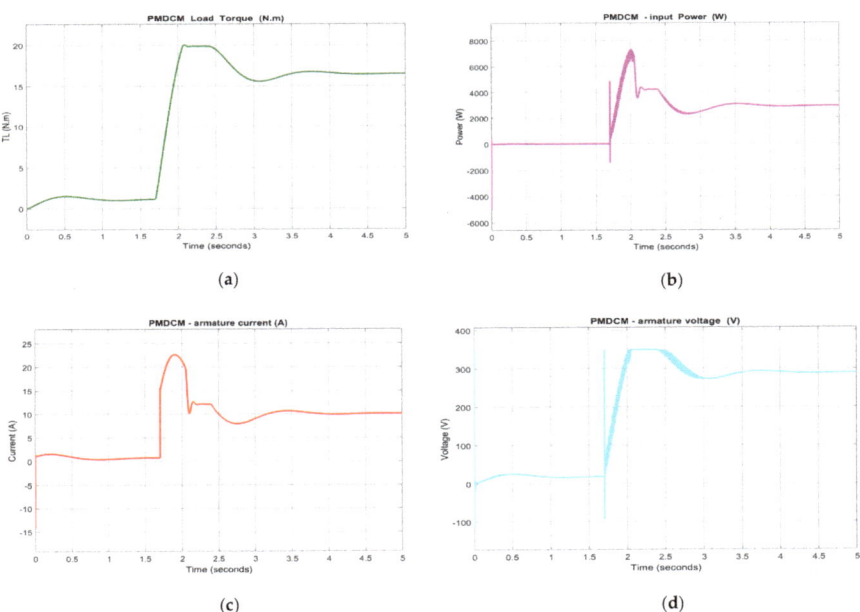

Figure 35. PMDCM—DC load. (**a**) PMDCM load torque; (**b**) PMDCM input absorbed power; (**c**) PMDCM armature current; (**d**) PMDCM armature voltage.

Figure 35a shows the linear evolution of the load torque with its PMDCM speed (scaling factor 0.011). Graph 35b describes the input power that changes abruptly in the step moment from 0 to 7 kW at t = 2 [s], followed by a decrease to 2.5 kW during steady state. Figure 35c shows the armature current absorbed by the PMDCM with a similar evolution trend as for the absorbed PMDCM power. Figure 35d shows the supplied armature voltage which is the same as the SC voltage, which justifies its presence to provide a large amount of energy, again protecting the Li-ion battery to make this effort. SC ensures the required voltage absorbed by PMDCM to achieve excellent speed profile tracking performance.

4. Li-Ion Battery Models Accuracy Performance—Battery Selection

4.1. Statistical Criteria to Asses the Accuracy of the Models

In a general formulation, for a better understanding of how to select an accurate Li-ion battery model, as well as a high-precision SOC state estimator and an excellent prediction of the selected battery output voltage, it can use some statistical criteria performance to compute the fitting errors between a set of candidates models reported in the literature [14,17,26,27]. Selection of Li-ion battery models and Kalman filter SOC estimators can be made by using the performance criteria developed in the "recent years in statistical learning, machine learning, and big data analytics" [27]. It is essential to emphasize the fact that now there are several criteria reported in the literature for models and estimators' selection, that "receives much attention due to growing areas in machine learning, data mining and data science" [27]. Among them, the mean squared error (MSE), root mean squared error (RMSE), R^2-squared, mean absolute squared error (MAE), standard deviation σ, the mean absolute percentage error (MAPE) [27], Adjusted R^2, Akaike's information criterion (AIC), Bayesian information criterion (BIC), AICc are "the most common criteria that have been used to measure model performance and select the best model from a set of potential models" [27].

Both models, i.e., the 3RC ECM and Simscape, are already validated by the available information extracted for each of them from the results of MATLAB simulations shown in Figure 13b, Figure 15, Figure 21a,b, that reveal an excellent accuracy due to SOC residual percentage errors being very low. A baseline for comparison is used the estimated value of ADVISOR SOC, as mentioned in the previous sections. The accuracy of both models is better compared to SOC residual error of 2%, usually reported in the literature for similar applications. The first model records a residual error of 1.2% (Figure 15) and the second one of 1.4% (Figure 21b).

Furthermore, to make a better delimitation between them, additional information is required. The values provided by all six statistical criteria, such as root mean squared error (RMSE), mean squared error (MSE), mean absolute error (MAE), standard deviation (std), mean absolute percentage error (MAPE) and squared, coefficient of determination (R^2-squared), as are defined in [26,27], is valuable information that makes the difference when two models perform close in terms of accuracy. These values are presented in Table 5, for the 3RC ECM, respectively Table 6, for Simscape Li-ion battery models. All these performance criteria have lower values thus validate, without any doubt, the both models. Moreover, because for both RMSE, MSE, std, MAPE models are very close and R^2-squared = 0.959 and 0.951, respectively, very close to 1, this is valuable information that indicates how close the values of the data set of the models are and of estimated values ADVISOR SOC. Thus, the overall performance is quite close, with a slight superiority of the 3RC ECR battery model, but the difference is still negligible..

Table 5. Statistical errors root mean squared error (RMSE), mean squared error (MSE), AMSE—Li-ion 3RC ECM SOC values versus ADVISOR SOC estimates.

RMSE	MSE	MAE	Standard Deviation (σ)	MAPE (%)	R^2-Squared
0.0075	0.00005	0.007	0.0384	1.12	0.959

Table 6. Statistical errors RMSE, MSE, MAE, std, MAPE and R^2 for Simscape Li-ion model validation versus ADVISOR estimate.

RMSE	MSE	MAE	Standard Deviation (σ)	MAPE (%)	R^2
0.0079	0.0000636	0.007	0.0384	1.19	0.951

4.2. Battery Selection Model

The lower values of all statistical performance criteria from Table 6, quite close to those from Table 5, justifies without doubts the validity of Simscape model. Based on the information collected from both Tables 5 and 6, a rigorous analysis reveals that the both models perform similarly and are suitable for building highly efficient and accurate SOC estimatiors in Part 2. However, an appropriate criterion to compare several candidate models is a hard task for any analyst since some criteria can be disadvantaged by the "model size of estimated parameters while the others could emphasis more on the sample size of a given data" [27]. As a general remark, it can be said that the Simscape Li-ion battery model excels in the following features: simplicity, friendly user-interface, and being fast to implement in real-time.

Also, the battery parameters can be extracted very easily for different chemistry and specifications, and the model is much more realistic in terms of the values of physical model parameters. However, a comparison of SOC performance, more relevant for highlighting the strengths and weaknesses of both models, is provided by the validation results of the MATLAB simulations shown in Figures 13b and 21a,b for the same FTP-75 driving cycle profile and equal SOC value estimated by the ADVISOR simulator 3.2. Following this test, excellent SOC accuracy is observed for both models. Valuable information on SOC accuracy is extracted from residual SOC errors generated in MATLAB and analyzed in Figures 15 and 21b. The results of the MATLAB simulations show a residual SOC error in the case of the 3RC ECM Li-ion battery model in the range [−1, 1, 0, 4], and the percentage of residual SOC error is below 1.2%. In contrast, for the Simscape model, the percentage of residual SOC error is less than 1.4%, compared to the typical value of 2% reported in the literature for similar applications. The features mentioned above and in addition to the SOC percentage of excellent residual error, much lower than the typical value of 2% reported in the literature, are sometimes even better, strongly recommending the choice of the Simscape model for a wide variety of HEV and EVs applications. In conclusion, these results are encouraging for the next step to develop the most suitable SOC estimators in Part 2. The results of the MATLAB simulations will confirm that all three SOC estimators work better for a design based on choosing the Simulink model, rather than adopting 3RC ECM battery mode, even if the accuracy performance of both battery models is equally sensitive.

5. Conclusions

In the current research paper, the following most relevant contributions of the authors can be highlighted:

- Selection for the same SAFT Li-ion battery of two models, the first model being a 3RC ECM and the second a highly non-linear MATLAB Simscape model, well known for its simplicity, excellent accuracy, practical value and suitable for real-time implementation.
- Model development in continuous and discrete-time state space representation.
- Validation of both models based on same FTP-75 driving cycle current profile test, using ADVISOR 3.2 software tool.
- Thermal model design and Simulink implementation.
- Opening of new research topics directions related to energy management systems, optimization techniques and HEV applications.

Based on six statistical criteria values for all three SOC estimators, as a behavior response to an FTP-75 driving cycle profile test, it was possible to decide based on SOC accuracy performance if both models are suitable to be used in Part 2 [30], for adaptive Kalman filter SOC estimators design and implementation. Furthermore, the overall performance analysis indicates that both models are accurate and suitable to be used in Part 2 [30]. In future work, our investigations will continue an improved modelling approach, by integrating the effect of degradation, temperature and SOC effects. New directions of research in energy management systems to develop power optimization techniques and for possible extensions to learning machine SOC estimation techniques will be a great challenge.

Author Contributions: R.-E.T., has contributed for, algorithm conceptualization, software, original draft preparation and writing it; N.T., has contributed for battery models investigation and validation, performed MATLAB simulations and formal analysis of the results; M.Z., has contributed for project administration, supervision, and results visualization; S.-M.R. has contributed for methodology, data curation and supervision. All authors have read and agreed to the published version of the manuscript.

Funding: This research received no external funding.

Acknowledgments: Research funding (discovery grant) for this project from the Natural Sciences and Engineering Research Council of Canada (NSERC) is gratefully acknowledged.

Conflicts of Interest: The authors declare no conflict of interest. This research received no external funding.

Abbreviations

EV	electric vehicle
HEV	hybrid electric vehicle
FCEV	fuel cell electric vehicle
HFCEV	hybrid fuel cell electric vehicle
PEM	polymer electrolyte membrane
SMACAR	small car
FC	fuel cell
UC	ultracapacitor
SC	supercapacitor
FCPEM	fuel cell polymer electrolyte membrane
EMS	energy management system
BMS	battery management system
ADVISOR	advanced vehicle simulator
EPA	environmental protection agency
UDDS	urban dynamometer driving schedule
FTP-75	Federal test procedure at 75 [degrees F]
RMSE	root mean squared error
MSE	mean squared error
MAE	mean absolute error
MAPE	mean absolute percentage error
R2/R	squared, coefficient of determination
std (σ)	standard deviation
OCV	open-circuit voltage
SOC	state of charge
NREL	National Renewable Energy Laboratory

Appendix A

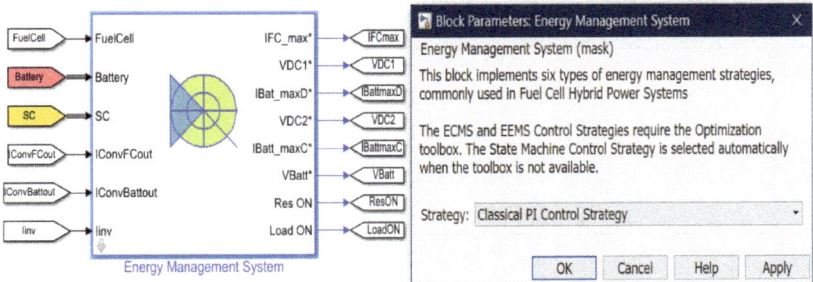

Figure A1. EMS—classical PI control strategy setup.

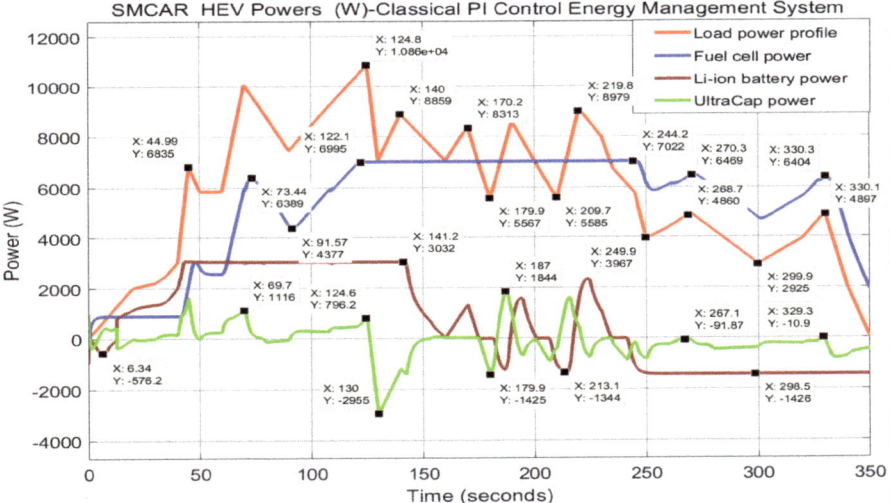

Figure A2. SMCAR HEV Powers—classical PI control EMS.

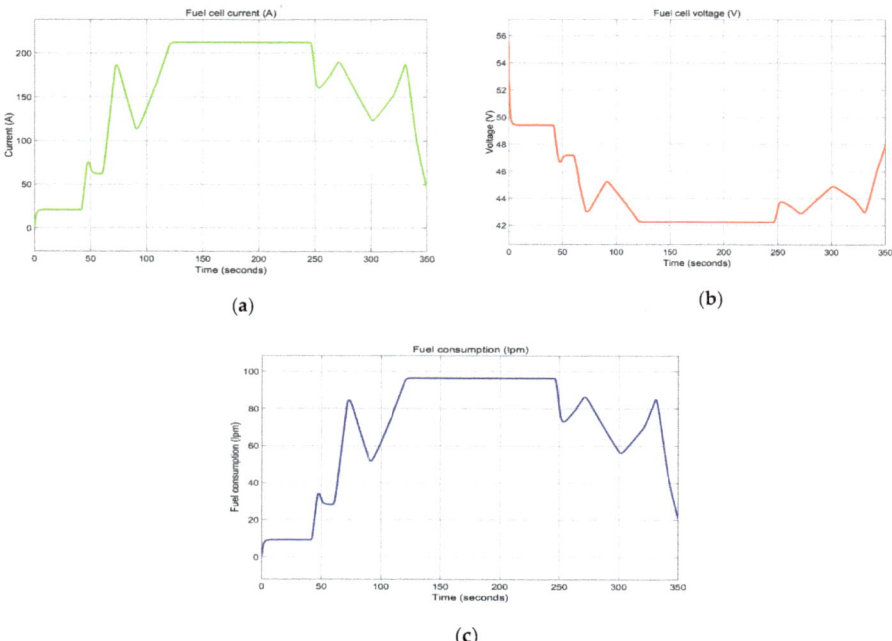

Figure A3. Fuel cell (FC). (**a**) FC current; (**b**) FC voltage; (**c**) FC consumption.

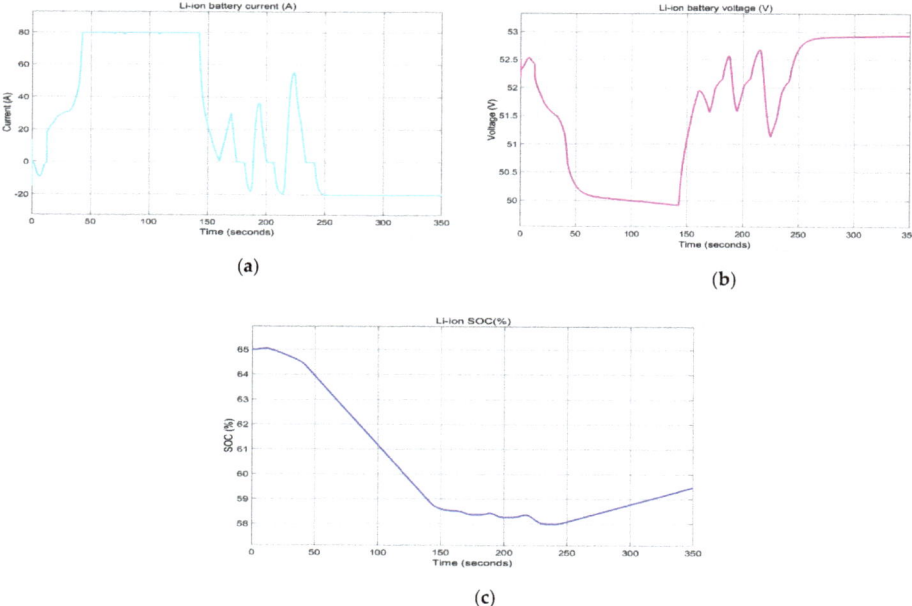

Figure A4. Li-ion battery specific variables. (**a**) Battery current; (**b**) battery voltage; (**c**) battery SOC.

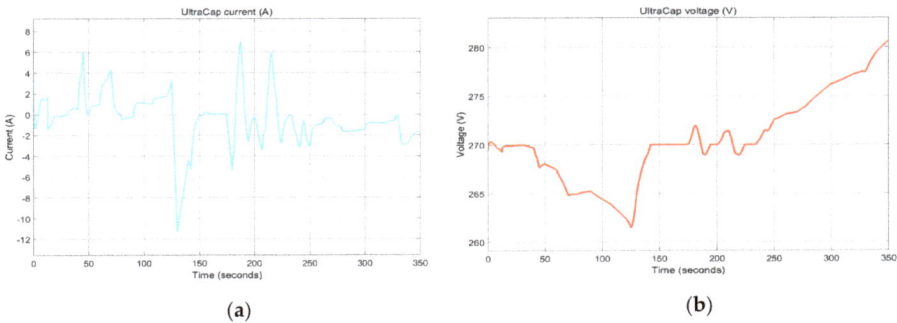

Figure A5. SC specific variables. (**a**) SC current variation; (**b**) SC voltage.

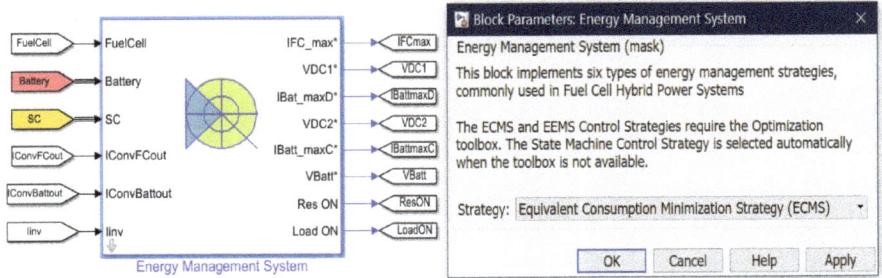

Figure A6. EMS-equivalent consumption minimization strategy setup.

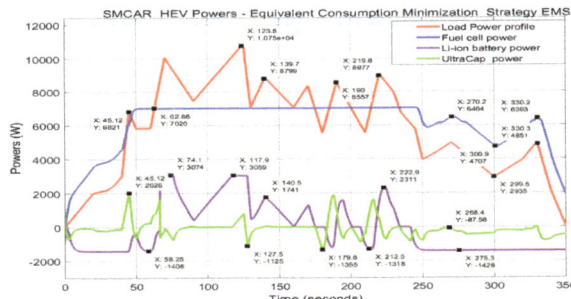

Figure A7. SMCAR HEV power-equivalent consumption minimization EMS.

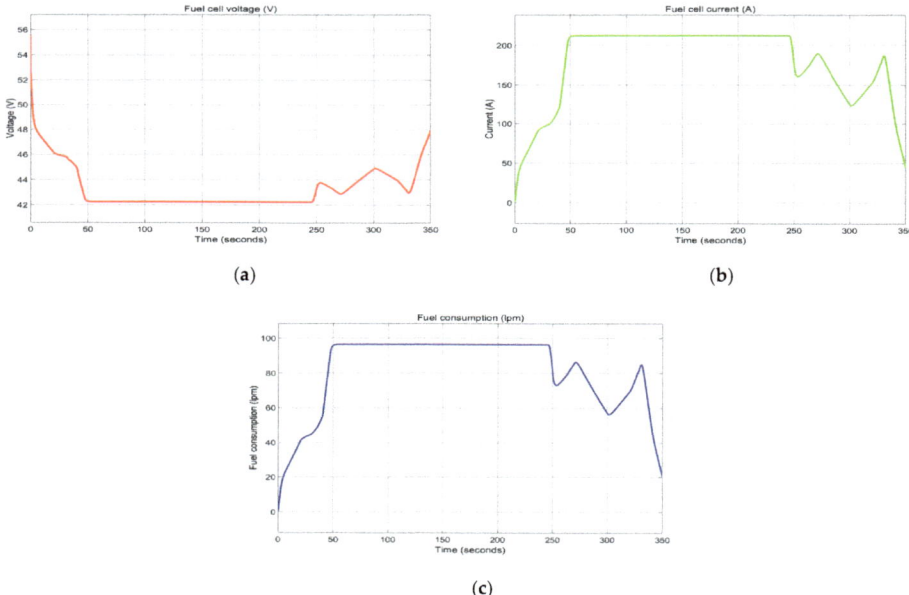

Figure A8. Fuel cell. (**a**) FC current; (**b**) FC voltage; (**c**) FC consumption.

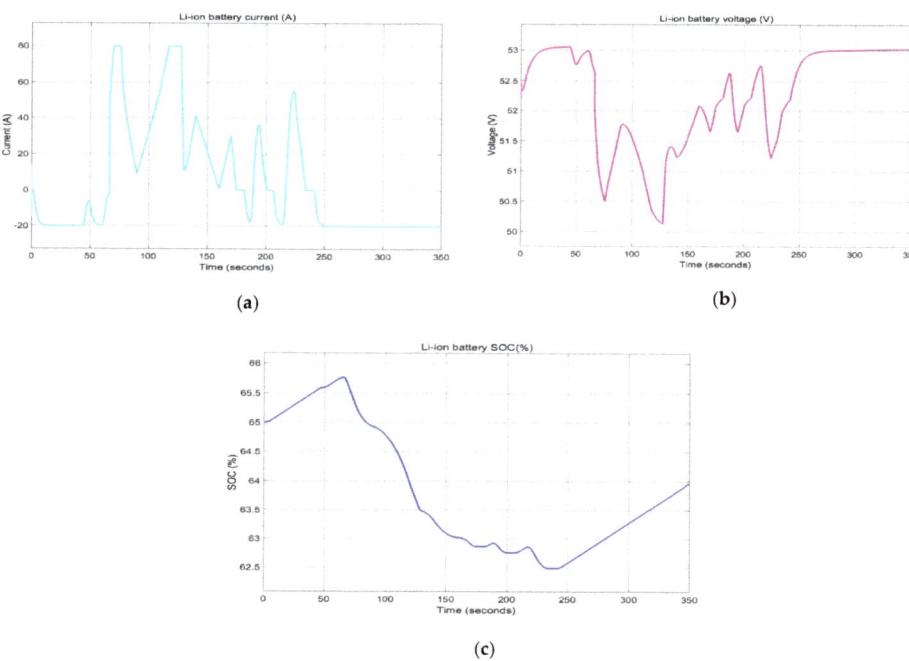

Figure A9. Li-ion battery specific variables. (**a**) Battery current; (**b**) battery voltage; (**c**) battery SOC.

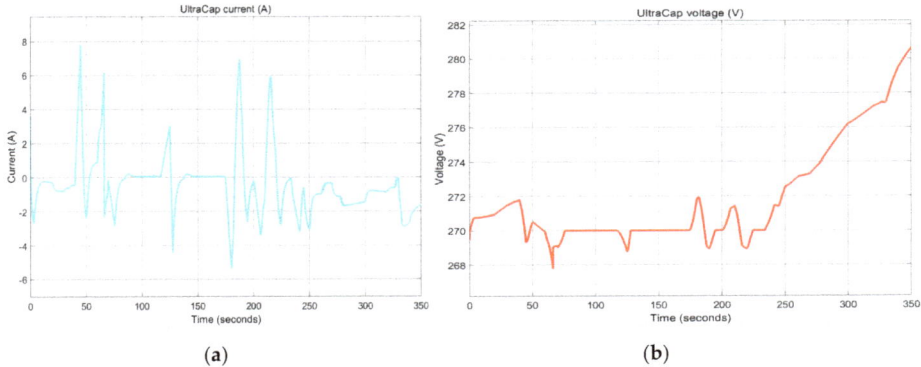

Figure A10. SC specific variables. (**a**) SC current variation; (**b**) SC voltage.

References

1. Alarco, J.; Talbot, P. The History and Development of Batteries. *Phys.org*. 30 April 2015. Available online: https://phys.org/news/2015-04-history-batteries.html (accessed on 28 June 2020).
2. Pressman, M. Understanding Tesla's Lithium Batteries. *Blog Post: Tesla News*. Available online: https://evannex.com/blogs/news/understanding-teslas-lithium-ion-batteries (accessed on 28 June 2020).
3. US Department of Energy. Energy Efficiency and Renewable Energy. Available online: https://afdc.energy.gov/vehicles/how-do-fuel-cell-electric-cars-work (accessed on 17 June 2020).
4. Energy.gov. Office of Energy Efficiency and Renewable Energy Fuel Cells. Available online: https://www.energy.gov/eere/fuelcells/fuel-cells (accessed on 28 June 2020).
5. Njoya, S.M. Design and Simulation of a Fuel Cell Hybrid Emergency Power System for a More Electric Aircraft: Evaluation of Energy Management Schemes. Ph.D. Thesis, University of Quebec from Montreal, Montreal, QC, Canada, 27 March 2013.
6. Jarushi, A.; Schofield, N. Modelling and Analysis of Energy Source Combinations for Electric Vehicles. EVS24 International Battery, Hybrid and Fuel Cell Electric Vehicle Symposium, Stavanger, Norway, May 13–16, 2009. *World Electr. Veh. J.* **2009**, *3*, 0796–0802. [CrossRef]
7. Njoya, S.M.; Dessaint, L.-A.; Liscouet-Hanke, S. Energy management Systems for a Hybrid Electric Source (Application for a More Electric Aircraft). MathWorks-MATLAB R2020a, Documentation. Available online: https://www.mathworks.com/help/physmod/sps/examples/energy-management-systems-for-a-hybrid-electric-source-application-for-a-more-electric-aircraft.html (accessed on 28 June 2020).
8. MathWorks-MATLAB R2020a, Documentation. Available online: https://www.mathworks.com/help/physmod/simscape/getting-started-with-simscape.html (accessed on 30 June 2020).
9. Burke, A.; Zhao, H. *Applications of Supercapacitors in Electric and Hybrid Vehicles*; Research Report-UCD-ITS-RR-15-09; Institute of Transportation Studies, University of California: Davis, CA, USA, 2015.
10. King, A. Power-Hungry Tesla Picks up Supercapacitor Maker. Royal Society of Chemistry, Chemistry World. 11 March 2019. Available online: https://www.chemistryworld.com/news/power-hungry-tesla-picks-up-supercapacitor-maker-/3010215.article (accessed on 29 June 2020).
11. Tejwani, V.; Suthar, B. Energy Management System in Fuel Cell, Ultracapacitor, Battery Hybrid Energy Storage. *World Acad. Sci. Eng. Technol. Int. J. Electr. Commun. Eng.* **2015**, *9*, 1492–1500.
12. Tudoroiu, R.-E.; Zaheeruddin, M.; Tudoroiu, N.; Budescu, D.D. Fuzzy Logic PID Control of a PMDCM Speed Connected to a 10-kW DC PV Array Microgrid—Case Study. In Proceedings of the Federated Conference on Computer Science and Information Systems (ACSIS), Berlin, Germany, 1–3 September 2019; Volume 18, pp. 359–362. [CrossRef]
13. Tudoroiu, R.-E.; Zaheeruddin, M.; Radu, S.-M.; Tudoroiu, N. A FLCPID RPM Hybrid Strategy for Speed Control-Case Study and Performance Comparison. In Proceedings of the 2020 2nd International Conference on Innovative Mechanisms for Industry Applications (ICIMIA), Bangalore, India, 5–7 March 2020. [CrossRef]

14. Tudoroiu, N.; Zaheeruddin, M.; Tudoroiu, R.-E. Real Time Design and Implementation of State of Charge Estimators for a Rechargeable Li-ion Cobalt Battery with Applicability in HEVs/EVs-A comparative Study. *Energies* **2020**, *13*, 2749. [CrossRef]
15. Keyser, M.A.; Pesaran, A.; Oweis, S.; Chagnon, G.; Ashtiani, C. Thermal Evaluation and Performance of High-Power Lithium-Ion Cells. Reprint. In Proceedings of the 16th Electric Vehicle Conference, Beijing, China, 1–3 October 1999; NREL: Golden, CO, USA. Available online: https://www.researchgate.net/publication/228834724 (accessed on 15 June 2020).
16. Brooker, A.; Haraldsson, K.; Hendricks, T.; Johnson, V.; Kelly, K.; Kramer, B.; Markel, T.; O'Keefe, M.; Sprik, S.; Wipke, K.; et al. *ADVISOR Advanced Vehicle Simulator*; Version 2003; Documentation-March 26, 2013; NREL: Golden, CO, USA, 10 October 2003. Available online: http://adv-vehicle-sim.sourceforge.net/ (accessed on 15 June 2020).
17. Tudoroiu, R.-E.; Zaheeruddin, M.; Radu, S.-M.; Tudoroiu, N. Real-Time Implementation of an Extended Kalman Filter and a PI Observer for State Estimation of Rechargeable Li-Ion Batteries in Hybrid Electric Vehicle Applications—A Case Study. *Batteries* **2018**, *4*, 19. [CrossRef]
18. Tudoroiu, R.-E.; Zaheeruddin, M.; Radu, S.M.; Tudoroiu, N. *New Trends in Electrical Vehicle Powertrains*; Martinez, L.R., Prieto, M.D., Eds.; BoD–Books on Demand: Norderstedt, Germany, 2019; Chapter 4. [CrossRef]
19. Farag, M. Lithium-Ion Batteries, Modeling and State of Charge Estimation. Master's Thesis, University of Hamilton, Hamilton, ON, Canada, 2013.
20. Plett, G.L. Extended Kalman filtering for battery management systems of LiPB-based HEV battery packs: Part 1. Background. *J. Power Sources* **2004**, *134*, 252–261. [CrossRef]
21. Plett, G.L. Extended Kalman filtering for battery management systems of LiPB-based HEV battery packs: Part 2. Modeling and identification. *J. Power Sources* **2004**, *134*, 262–276. [CrossRef]
22. Plett, G.L. Extended Kalman filtering for battery management systems of LiPB-based HEV battery packs: Part 3. State and parameter estimation. *J. Power Sources* **2004**, *134*, 277–292. [CrossRef]
23. Tudoroiu, N.; Radu, S.M.; Tudoroiu, R.-E. *Improving Nonlinear State Estimation Techniques by Hybrid. Structures*, 1st ed.; LAMBERT Academic Publishing: Saarbrucken, Germany, 2017; p. 56. ISBN 978-3-330-04418-0.
24. Hussein, A.A.; Batarseh, I. An overview of generic battery models. In Proceedings of the 2011 IEEE Power and Energy Society General Meeting Conference, Detroit, MI, USA, 24–28 July 2011. [CrossRef]
25. Geetha, A.; Subramani, C. A significant energy management control strategy for a hybrid source EV. *Int. J. Electr. Comput. Eng.* **2019**, *9*, 4580. [CrossRef]
26. Wikipedia. Mean Absolute Percentage Error. Available online: https://en.wikipedia.org/wiki/Mean_absolute_percentage_error (accessed on 6 July 2020).
27. Pham, H. A New Criterion for Model Selection. *Mathematics* **2019**, *7*, 1215. [CrossRef]
28. Zhang, R.; Xia, B.; Li, B.; Cao, L.; Lai, Y.; Zheng, W.; Wang, H.; Wang, W. State of the Art of Li-ion Battery SOC Estimation for Electrical Vehicles. *Energies* **2018**, *11*, 1820. [CrossRef]
29. Kim, T. A Hybrid Battery Model Capable of Capturing Dynamic Circuit Characteristics and Nonlinear Capacity Effects. Master's Thesis, University of Nebraska-Lincoln, Lincoln, NE, USA, 2012.
30. Tudoroiu, R.-E.; Zaheeruddin, M.; Tudoroiu, N.; Radu, S.M. SOC Estimation of a Rechargeable Li-Ion Battery in fuel-Cell Hybrid Electric Vehicles-Comparative Study of Accuracy and Robustness performance Based on Statistical Criteria. Part II: SOC Estimators. *Batteries* **2020**, *6*. in press.

© 2020 by the authors. Licensee MDPI, Basel, Switzerland. This article is an open access article distributed under the terms and conditions of the Creative Commons Attribution (CC BY) license (http://creativecommons.org/licenses/by/4.0/).

Article

SOC Estimation of a Rechargeable Li-Ion Battery Used in Fuel Cell Hybrid Electric Vehicles—Comparative Study of Accuracy and Robustness Performance Based on Statistical Criteria. Part II: SOC Estimators

Roxana-Elena Tudoroiu [1], Mohammed Zaheeruddin [2], Nicolae Tudoroiu [3,*] and Sorin-Mihai Radu [4]

[1] Department of Mathematics and Informatics, University of Petrosani, 332006 Petrosani, Romania; tudelena@mail.com
[2] Department of Building, Civil and Environmental Engineering, University Concordia from Montreal, Montreal, QC H3G 1M8, Canada; zaheer@encs.concordia.ca
[3] Department of Engineering Technologies, John Abbott College, Saint-Anne-de-Bellevue, QC H9X 3L9, Canada
[4] Department of Control, Computers, Electrical and Power Engineering, University of Petrosani, 332006 Petrosani, Romania; sorin_mihai_radu@yahoo.com
* Correspondence: ntudoroiu@gmail.com; Tel.: +1-514-966-5637

Received: 4 August 2020; Accepted: 13 August 2020; Published: 14 August 2020

Abstract: The purpose of this paper is to analyze the accuracy of three state of charge (SOC) estimators of a rechargeable Li-ion SAFT battery based on two accurate Li-ion battery models, namely a linear RC equivalent electrical circuit (ECM) and a nonlinear Simscape generic model, developed in Part 1. The battery SOC of both Li-ion battery models is estimated using a linearized adaptive extended Kalman filter (AEKF), a nonlinear adaptive unscented Kalman filter (AUKF) and a nonlinear and non-Gaussian particle filter estimator (PFE). The result of MATLAB simulations shows the efficiency of all three SOC estimators, especially AEKF, followed in order of decreasing performance by AUKF and PFE. Besides, this result reveals a slight superiority of the SOC estimation accuracy when using the Simscape model for SOC estimator design. Overall, the performance of all three SOC estimators in terms of accuracy, convergence of response speed and robustness is excellent and is comparable to state of the art SOC estimation methods.

Keywords: SAFT lithium-ion battery; Simscape model; 3RC ECM Li-ion battery model; state of charge; adaptive EKF SOC estimator; adaptive UKF SOC estimator; particle filter SOC estimator

1. Introduction

In recent years, the lithium-ion battery has proven to be an ideal safety battery for hybrid electric vehicles, with high discharge power, environmental protection, low pollution, and long life [1–19]. Some details about its features, modelling, and hybrid combinations with different power sources in a fuel cell electric vehicle (FCEV) and power distribution controlled and optimized by an energy management system (EMS) are shown in Part 1 [20]. It is worth mentioning that the battery SOC is an essential internal parameter that is continuously monitored by a battery management system (BMS) to prevent dangerous situations and improves battery performance. The Li-ion battery as a direct energy supply and its SOC have a significant impact on the HEV's performance. Besides, the amount of SOC is crucial for safe operation of the Li-ion battery and its prolongation of life, so an accurate estimate of the SOC has an important theoretical significance and application value [1,2,5,6].

Typically, for calculation based on the coulomb counting method, the SOC is "tracking according to the discharging current" [5,6,12]. In the absence of a measurement sensor, the SOC cannot be measured directly; thus, its estimation using a Kalman filter technique is required [5].

Typically, the Kalman filter SOC estimators are model based, so both battery models—the linear RC equivalent electrical circuit (ECM) and the nonlinear Simscape generic model developed and analyzed in Part 1 [20] are beneficial for designing and implementing a high accuracy SOC estimator [1–10,12–19]. For better documentation and information for the reader, the diagrams of both models represented in Part 1 [20] are taken over and repeated in Appendix A.1, Figure A1a–c. The Li-ion battery is an essential component of the battery management system (BMS) that plays an important role for improving the battery performance [2,5,6,12]. More details on the definition, the role, the main components (hardware and software) and the multitask functions can be found in [2,5]. Motivated by the results obtained in Part 1 [20], this article focuses on the design and implementation of three real-time SOC estimators on a MATLAB R2020a simulation environment. The remaining sections of this paper are structured as follows. Section 2 makes a presentation of state of the art of Li-ion battery SOC estimation Kalman filter techniques. Section 3 describes three of the most suitable SOC estimators in HEV applications and for each estimator shows the MATLAB simulation results. Section 4 analyses, for each SOC estimator, the SOC accuracy, convergence speed and robustness performance using six statistical criteria, defined in Part 1 [20]. Section 5 highlights the authors' contributions to this research paper.

2. State of the Art of Li-Ion Battery SOC Estimation Kalman Filter Techniques

The most popular nowadays, Kalman filter (KF) is the "optimum state estimator and intelligent tool for a linear system", beneficial for estimating the Li-ion battery dynamic states and parameters [8].

Its "predictor–corrector" structure, more precisely the "self-correcting" nature, is the most attractive feature of the KF algorithm when the system is running, which helps to "tolerate large variations" in the estimated SOC values, as mentioned in [6]. Besides, it can significantly improve the "accuracy and robustness of battery SOC estimation", as well as the filtering of noise that realistically occurs in the measurement output dataset and the battery model process. The accuracy, response speed convergence, robustness, and noise filtration, in the proposed case study, are approached in some detail in Sections 4.1–4.5. The values of statistical criteria from Tables 1 and A1, Tables A2–A4 analyzed in Section 4.5, play an essential role in the analysis of SOC estimation performance for all three SOC estimators and both models of the Li-ion battery. It is worth mentioning that all the Kalman filter state estimators played a crucial role in the last six decades, reforming the whole theory of automatic control systems, both theoretically and in terms of applicability. A combination of the KF state estimator and the Ah Coulomb counting method can be used to "compensate for the non-ideal factors that can prolong the operation of the battery" [6]. However, there are situations when some Li-ion battery models have a dynamic that is "extremely nonlinear" and therefore "the linearization error may occur due to the lack of precision in the extension of the first series Taylor series in extremely nonlinear conditions" [5]. The simplicity of the SOC EKF estimator design and real-time MATLAB implementation is among two main features that motivated many researchers to apply it to a variety of Li-ion battery models, as in [2,3,6–9]. A new state of the art analysis on Li-ion BMSs is presented in [12], which includes a brief overview presentation of the most common adaptive filtration techniques for SOC estimation reported in the literature. Similarly, in [6], the authors present an interesting state of the art study on SOC estimation of the Li-ion battery for electric vehicles, in which an entire subsection examines all existing adaptive SOC filtration estimation techniques reported in the literature. A brief review on SOC estimating techniques related to Li-based batteries can be found in [13]. In [14], a new approach, the dual EKF SOC estimator of first-order RC ECM Li-ion battery model state and parameters, is well documented. The SOC simulations resulting from research paper [14] reveal excellent accuracy for SOC estimation.

Table 1. Statistical criteria—state of charge (SOC) estimators (default value SOCini = 0.7).

Performance	Li-Ion Battery 3RC ECM, $\sigma = 0.03713$				Li-Ion Battery Simulink Simscape Model, $\sigma = 0.036248$			
	ADV	AEKF	AUKF	PFE	ADV	AEKF	AUKF	PFE
RMSE	0.0075	0.007	0.0052	0.02398	0.0079	0.0037	0.0135	0.0084
MSE	0.00005	0.000049	2.6×10^{-5}	0.0005	6×10^{-5}	1.4×10^{-7}	0.00018	7×10^{-5}
MAE	0.0070	0.0051	0.0059	0.0179	0.0075	0.000214	0.0127	0.0065
Standard deviation (σ)	0.0384	0.043	0.0369	0.0554	0.0384	0.036242	0.044	0.0358
MAPE (%)	1.1249	0.849	0.7972	1.08	1.1965	0.50	2.178	1.06
R^2	0.9591	0.864	0.9805	0.679	0.9515	0.999	0.908	0.946
Result Hierarchy		2	1	3		1	3	2

First place: 1; second place: 2; third place: 3.

Still, the robustness of the algorithm in [14] is lacking; it is strengthened in our research for five different scenarios and two battery models. Besides this, six performance analysis criteria are defined and used to assess the accuracy and robustness of SOCs. On the other hand, the authors of [14], in a new frame of a fault detection and isolation (FDI) approach, develop an SOC AEKF estimator for a Li-ion battery. A rigorous analysis of fault estimation performance, injected into BMS current and voltage sensor, showed a high accuracy and robustness to a 20% initial initialization of SOC error for an urban dynamometer driving schedule (UDDS) driving cycle profile test. The SOC accuracy and robustness performance are comparable to those obtained in our research for 30% initialization SOC error (scenario R1 for both battery models and each SOC estimator) and for an FTP-75 driving cycle profile test that includes the UDDS in the first 1379 s. Of course, to analyze the impact of each fault on the SOC estimation performance it was beneficial to see the fault SOC estimated values. An AEKF fading (AFEKF) approach is proposed in [15] for the accuracy of Li-ion battery SOC estimation and the convergence rate, which can reduce the SOC estimation error to less than 2%. The AFEKF SOC estimator performs better in terms of accuracy, robustness and convergence speed for a 20% initialization SOC error, but in our research similar performances are obtained at the initialization of 30% and 50% SOC errors (scenario R1), and also in combination with capacity degradation (scenario R2), noise level change (scenario R3) and the effects of temperature on the internal resistance of the battery (scenario R4). The result of the MATLAB simulations reveals that the AEKF SOC estimator works successfully in all five scenarios, especially for the Simscape model.

A similar situation is reported in the literature, in reference [16], where the authors investigate an RC ECM Li-ion battery model, and the SOC accuracy performance and robustness are analyzed for 20% initialization of SOC error. The SOC estimated error of the AEKF SOC estimator is more significant than 2% during the steady-state for a considerable window length $t \in (800, 2200)$ s of SOC residual [16] compared to the AEKF SOC estimator used in our research for which the SOC estimated error is 0.32% for third order resistor capacitor (3RC) equivalent circuit model (ECM), if a 20% initialization SOC error (SOCini = 50%) is under investigation. For performance comparison purposes, Figure 1 shows a complete picture of Li-ion battery AEKF SOC, i.e., accuracy and robustness performance, for a 20% initialization SOC error, such as is reported in several references in the literature. Typically, in our research, for MATLAB simulations, 30% and 50% initialization of SOC errors, in combination with different scenarios, are under consideration.

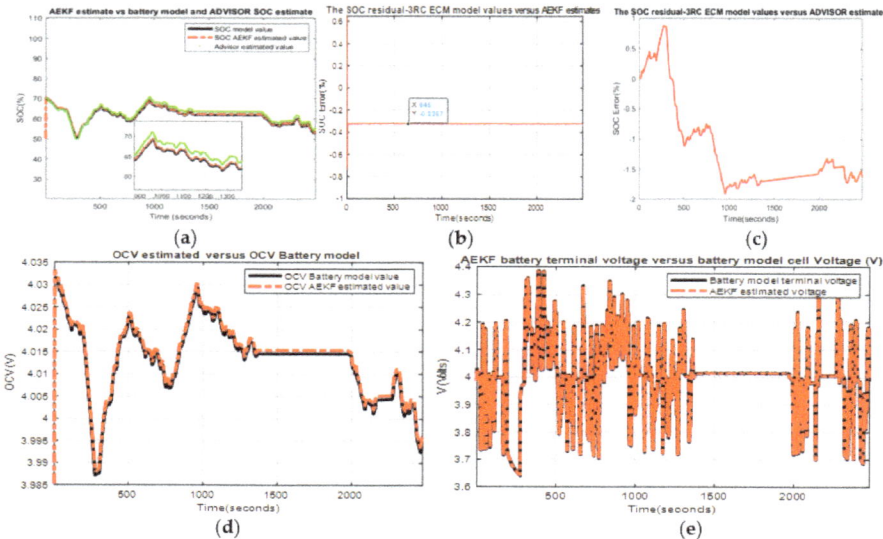

Figure 1. The Li-ion battery adaptive extended Kalman filter (AEKF) SOC and robustness performance for an initializing 20% SOC error (SOCini = 50%; SOCini for advanced vehicle simulator ADVISOR and battery model is 70%) (**a**) Li-ion battery model SOC (blue), ADVISOR SOC estimate (green), AEKF SOC (red); (**b**) SOC residual error AEKF vs ADVISOR SOC estimate; (**c**) SOC residual error AEKF vs battery model; (**d**) AEKF open-circuit voltage (OCV) vs battery model OCV (true value); (**e**) AEKF battery terminal output voltage vs battery model terminal output voltage (true value).

For a similar RC ECM battery model, reference [17] shows the SOC accuracy and robustness of the AEKF SOC estimator for a UDDS current profile test, and 20% initialization SOC error. The simulation results in [17] reveal that the estimated SOC error reaches, during the UDDS driving cycle test, even 5%. In reference [18], based on an RC ECM model, an adaptive Kalman filter (AKF) is implemented and the SOC is set for initialization to SOCini = 76% and SOCini = 81%, compared to the default value SOCini = 80%. During the driving cycle the SOC errors reach 5% for first case and 4% for the second case. The selective results reported in the literature, highlighted in this paragraph, are significant for demonstrating the effectiveness of all three Kalman filter SOC estimators. In conclusion, comparing the simulation results obtained in our research work with those reported in the literature and mentioned above, it can be affirmed that the proposed SOC estimators and both models of Li-ion batteries are very efficient and work very well. The AEKF SOC estimator was chosen as a baseline because its results have a slight superiority compared to the other two competitors, namely the AUKF and PFE SOC estimators. AEKF proved to be a strong competitor compared to many other SOC estimators reported in the literature. In general, it can be said with confidence that the SOC AEKF estimator adopted in the present research with a correct design and with the parameters established at appropriate optimal values has better results than those found in the literature. Fundamental work related to the unscented Kalman filter (UKF) estimator is outlined in [7], which provides a strong theoretical background. Moreover, a particle filter estimator (FPE) is used to estimate the states, estimating the "probability density function" of a nonlinear dynamics of the Li-ion battery model, using a Monte-Carlo simulation technique, such as is developed in [11].

3. Li-Ion Battery SOC—Adaptive and Particle Filter Estimators

In this section, an overview of two Kalman filter SOC estimators with adaptive function is provided, namely a linearized adaptive Kalman filter (AEKF) [5,6,15–19] and an unscented adaptive

Kalman filter (AUKF). A successful implementation of both SOC estimators is performed on the software platform MATLAB R2020a, which estimates the SOC of a Li-ion SAFT battery with a rated capacity of 6 Ah and a nominal voltage of 3.6 V. Both SOC estimators under investigation are model based; thus, a dynamic state space representation model of the Li-ion battery is required in order to develop a simulation model for the emulation of nonlinear battery behavior [5,6]. In the case study, the set of equations that describe both models developed in Part 1 [20] is used, namely a 3RC ECM Li-ion battery model and a Li-ion battery Simscape model.

3.1. Adaptive Extended Kalman Filter (AEKF) Overview Presentation

The AEKF SOC estimator is a standard EKF, such as those developed in [2,5,6,15–19], with improved performance by using a memory fading factor [16] or adaptive correction of process and measurement noise covariances [6]. In [5], the SOC AEKF estimator combines both memory fading and noise correction. Encouraged by the preliminary results obtained in [5], the current research paper implements the same version of the AEKF SOC estimator adapted for each of the Li-ion battery models under investigation, namely the 3RC ECM and Simscape models. In the following are underlined only some interesting implementation aspects related to the AEKF estimation algorithm. The AEKF algorithm can improve SOC estimation performance by using "a fading memory factor to increase the adaptiveness for the modelling errors and the uncertainty of Li-ion battery SOC estimation, as well as to give more credibility to the measurements" [19]. It is based on the linearized models of the Li-ion battery described in the previous section. An excellent feature of the AEKF SOC estimator is that it is easy to implement in real time, due to its "recursive predictor–corrector structure that allows the time and measurement updates at each iteration" [5,19].

The tuning parameters of the AEKF SOC estimator are the following: $Q(0)$ and $R(0)$, $\hat{P}(0) = \hat{P}(0|0)$, the fading factor α and the window length L, obtained by a "trial and error" procedure based on the designer's empirical experience.

For simulation purposes, to test the effectiveness of the AEKF SOC estimator, the Kalman filter estimator parameters are set up for an FTP = 75 driving cycle profile test to the following set of values, $Q(0) = 5 \times 10^{-4}$, $R(0) = 0.4 \times 10^{-4}$, $\alpha = 0.9$, $\hat{P}(0) = 10^{-10}$, $L = 100$ samples for the Simscape battery model, and a second set of values $Q(0) = diag([qw\ qw\ qw\ qwSOC])$, $qw = 2 \times 10^{-3}$, $qwSOC = 0.5$, $R(0) = 0.02$, $\alpha = 0.9$, $\hat{P}(0) = 10^{-10}$, $L = 80$ samples for the 3RC ECM Li-ion battery model.

The MATLAB simulation results for an FTP-75 driving cycle current profile test are shown for all three SOC estimators, adapted for each Li-ion battery model described in the previous section and for the following five main scenarios, defined as:

- Scenario R0—SOC estimator accuracy based on the SOC residual curve, for an SOC initial value SOCini = 70% (i.e., same as the advanced vehicle simulator (ADVISOR) SOC estimated value) and the statistical criteria values, i.e., RMSE, MSE, MAE, std, MAPE and R-squared, given in Table 1.
- Scenario R1—SOC estimator robustness to changes in initial SOC value, SOCini = 0.4. The MATLAB simulation results are shown in Appendix A.1, and the statistical criteria values are provided in Appendix A.2, Table A1.
- Scenario R2—SOC estimator robustness to simultaneous changes, i.e., SOCini = 1 and battery ageing effects (a decrease in battery capacity by 30%, i.e., Qnom decreases from 6 Ah to 4.2 Ah). The MATLAB simulation results are shown in the main part of the manuscript, and statistical criteria values are given in Appendix A.2, Table A2.
- Scenario R3—SOC estimator robustness to simultaneous changes, namely in SOCini (SOCini = 0.4) and to 10 times increase in measurement noise level (e.g., σ = 0.01). The MATLAB simulation results are shown in Appendix A.1, and the statistical criteria values are given in Appendix A.2, Table A3.
- Scenario R4—SOC estimator robustness to simultaneous changes, such as in SOCini (SOCini = 0.2), temperature effects on internal resistance Rin and polarization constant Kp (only for Simscape

Li-ion battery model) to changes in ambient temperature ($T_0 = 293.15$ K, equivalent to 20 °C), as is shown in Part 1 [20], p. 12 for thermal model.

The MATLAB simulation results are depicted in the main part of the manuscript and in Appendix A.1, and the statistical criteria values are given in Appendix A.2, Table A4.

For each SOC estimator, these abbreviations of the five scenarios in the following text inserted into the manuscript are used to avoid repeating the words.

3.1.1. MATLAB Simulation Results for 3RC ECM Battery Model—Accuracy and Robustness Scenarios

- Scenario R0. The MATLAB simulation results for this scenario are shown in Appendix A.1, Figure A12a–c, and the statistical criteria values are given in Table 1.

  Performance analysis:

  ○ SOC of high accuracy and a great battery output voltage prediction.
  ○ The residual error is quite close to 1.5%, which is comparable to the results reported in the literature.

- Scenario R1. The MATLAB simulation results for the first scenario are presented in Appendix A.1, Figure A13a–c, and the statistical criteria values are given in Appendix A.2, Table A1.

  Performance analysis:

  ○ The simulation results reveal excellent SOC accuracy and a great robustness to changes in the initial SOC value.
  ○ The steady-state residual error is quite close to zero, which is an excellent result.

- Scenario R2. The MATLAB simulation results for the second scenario are depicted in Figure 2 and statistical criteria values are given in Appendix A.2, Table A2.

  Performance analysis:

  ○ The SOC accuracy is good and the robustness to ageing effects is great.
  ○ The steady-state residual error converges to −2%, which is a good result.

- Scenario R3. The MATLAB simulation results for the third scenario are shown in Appendix A.1, Figure A14a–c, and the statistical criteria values are given in Appendix A.2, Table A3.

  Performance analysis:

  ○ The SOC accuracy is bad and the robustness to an increased noise level is bad.
  ○ The steady-state residual error converges to −11%, which is a bad performance.

- Scenario R4: The MATLAB simulation results for the fourth scenario are shown in Figure 3 and the statistical criteria values are given in Appendix A.2, Table A4.

  Performance analysis:

  ○ The SOC accuracy is bad and the robustness to temperature effects is bad.
  ○ The steady-state residual error converges to −18%, which is a bad performance.

Batteries **2020**, *6*, 41

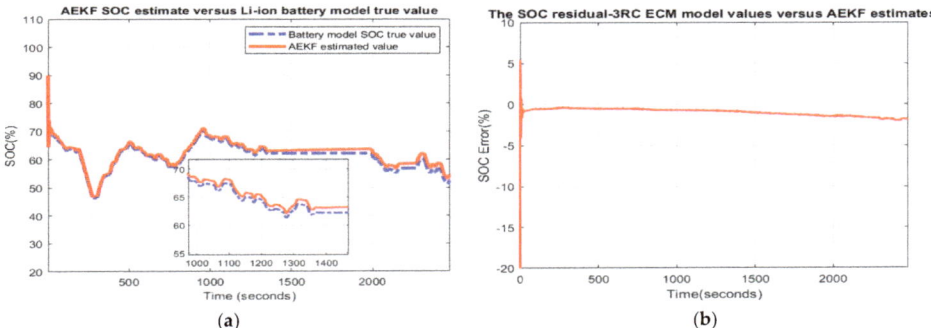

Figure 2. Robustness to simultaneous changes, SOCini = 1, Qnom = 4.2 Ah (ageing effects); (**a**) AEKF SOC value versus battery model true value; (**b**) SOC residual.

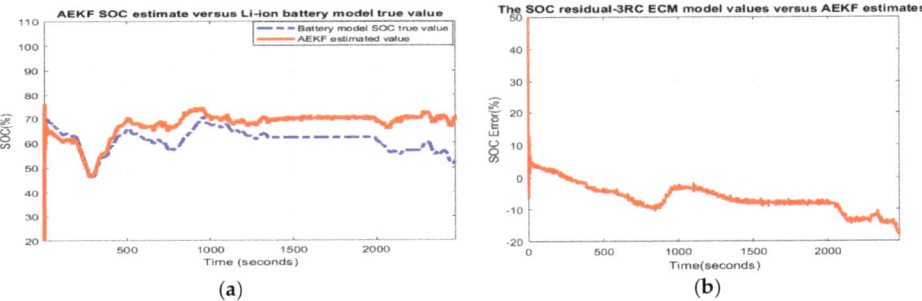

Figure 3. Robustness to simultaneous changes, SOCini = 0.2, and output temperature profile changes; (**a**) AEKF SOC value versus battery model true value; (**b**) SOC residual.

3.1.2. MATLAB Implementation and Simulation Results for Simulink Simscape Battery Model—Accuracy and Robustness Scenarios

- Scenario R0. The MATLAB simulation results for this scenario are depicted in Appendix A.1, Figure A15a–c, and the statistical criteria values are given in Table 1.

    Performance analysis:

    ○   SOC accuracy is excellent and battery output voltage prediction is great.
    ○   The residual error is quite close to 0.4%, which is very good result.

- Scenario R1. The MATLAB simulation results for first scenario are exposed in Appendix A.1, Figure A16a–c, and the statistical criteria values are given in Appendix A.2, Table A1.

    Performance analysis:

    ○   SOC accuracy is excellent and the robustness to changes in the SOCini is great.
    ○   The residual error has some variations near the origin but is quite close to zero in steady state, which is very good result.

- Scenario R2. The MATLAB simulation results for the second scenario are visible in Figure 4 and statistical criteria values are given in Appendix A.2, Table A2.

    Performance analysis:

    ○   SOC accuracy is good and robustness to ageing effects is great.

○ The residual error is quite close to 2% in steady state, which is a good result
- Scenario R3. The MATLAB simulation results for the third scenario are shown in Appendix A.1, Figure A17a–c, and the statistical criteria values are given in Appendix A.2, Table A3.

Performance analysis:

○ SOC accuracy is good and robustness to noise level is great.
○ The residual error has some variations near origin and in steady state it is quite close to 2%, which is a good result.

- Scenario R4. The MATLAB simulation results for the fourth scenario are depicted in Figure 5 and the statistical criteria values are given in Appendix A.2, Table A4.

Performance analysis:

○ SOC accuracy is good and robustness to temperature effects is great.
○ The residual error is quite close to 0% in steady state, which is an excellent result.

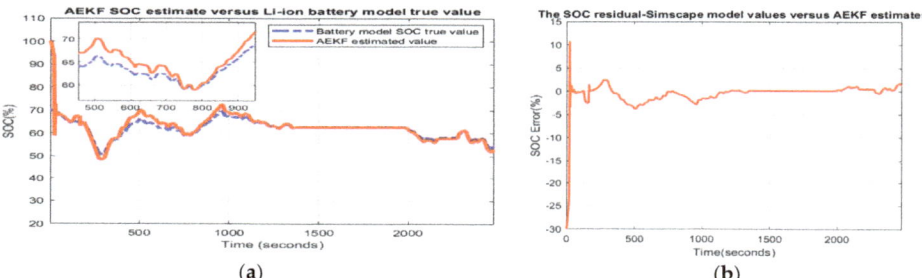

Figure 4. Robustness to simultaneous changes, SOCini = 1, Qnom = 4.2 Ah (ageing effects); (**a**) AEKF SOC value versus battery model true value; (**b**) SOC residual.

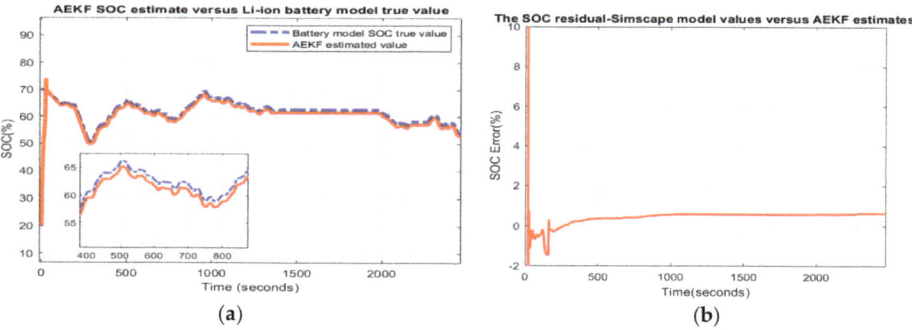

Figure 5. Robustness to simultaneous changes, SOCini = 0.2, and output temperature profile changes; (**a**) AEKF SOC value versus battery model true value; (**b**) SOC residual.

Roughly, based on the MATLAB simulation results of the AEKF SOC performance obtained for each model, both the 3RC ECM and Simscape models, it seems that the AEKF SOC estimator works better in all five scenarios based on the Simscape model.

3.2. Adaptive Unscented Kalman Filter (AUKF)

The AUKF SOC estimator [4,6,8,9] is an extremely precise algorithm, suitable for nonlinear dynamics of Li-ion battery models, compared to the AEKF SOC estimator [5,6,15–19], which only deals with models that have a linearized dynamic, so the calculation of Jacobian matrices is required, which is time consuming.

In the following, the steps of a general formulation of the AUKF SOC estimator that can be easily adapted to each model under investigation are presented briefly.

A standard UKF estimator is today one of the most popular estimators for states and nonlinear process parameters reported in the literature [7,12]. The AUKF SOC estimator adopted for the current research paper has the same steps as in [6]; our contribution is the adaptation of the algorithm to both the proposed Li-ion battery models, described in the previous section, and the parameter adjustment procedure for achieving an excellent accuracy for SOC.

UAKF SOC estimator algorithm steps [5,6]:

[UAKF1]. Write the battery model equations in discrete time state space representation.

1.1 The Simscape model is given by a set of two equations described in Part 1 [20], p. 20.
1.2 The 3RC ECM is given by a compact set in a matrix representation, as is shown in Part 1 [20], p. 12.
1.3 Model general formulation:

$x_{k+1} = f(x_k, u_k) + q_{w,k}$.
$y_k = g(x_k, x_k) + r_{v,k}$.

[UAKF2]. Initialization.

For $k = 0$, let:

$\hat{x}_0^+ = E[x_0]$ denotes the mean of initial value (predicted value).

$P_{x,0}^+ = E[(x_0 - \hat{x}_0^+)(x_0 - \hat{x}_0^+)^T]$ is the state covariance matrix (predicted state).

[UAKF3]. Computation.

3.1 Generate sigma points and weighting coefficients at time $k-1$, $k = \overline{[1, \infty]}$

$\widetilde{X}_0 = \hat{X}_{k-1}$ is the mean of the state at time $k-1$.
$\widetilde{X}_{k-1}^i = \hat{X}_{k-1} + \left(\sqrt{(n+\lambda)P_{k-1}}\right)_i$, $i = \overline{0, n}$ represent the sigma points.
$\widetilde{X}_{k-1}^j = \hat{X}_{k-1} + \left(\sqrt{(n+\lambda)P_{k-1}}\right)_{j-n}$, $j = \overline{n+1, 2n}$ are the sigma points.
$W_m^{(0)} = \frac{\lambda}{n+\lambda}$ — mean weights.
$W_c^{(0)} = \frac{\lambda}{n+\lambda} + 1 - \alpha^2 + \beta$ denote the covariance weights.
$W_m^{(i)} = W_c^{(i)} = \frac{\lambda}{2(n+\lambda)}$ are the mean and covariance weights.

[UAKF4]. Prediction phase (Forecast):

For $k = \overline{[1, n]}$, compute:

4.1 State estimate time update:

$\widetilde{X}_{k|k-1}^i = f(\hat{X}_{k|k-1}^i, u_k)$ is the prediction state vector (passing sigma points through function f (.,.)).
$\hat{x}_{k-1} = \sum_{i=0}^{2n} W_m^{(i)} \widetilde{X}_{k|k-1}^i$ designates the state estimate at time $k - 1$.
$P_{x,k|k-1}^- = \sum_{i=0}^{2n} W_c^{(i)} [\widetilde{X}_{k|k-1}^i - \hat{x}_{k-1}][\widetilde{X}_{k|k-1}^i - \hat{x}_{k-1}]^T$ denotes the prediction covariance matrix.
$Y_{k|k-1}^i = g(\hat{X}_{k|k-1}^i, u_k)$ are the output sigma points (passing sigma points through output function g(.,.)).
$\hat{y}_{k-1} = \sum_{i=0}^{2n} W_m^{(i)} Y_{k|k-1}^i$ is the output mean estimate.

[UAKF5]. Correction update phase (analysis):

5.1 Update the covariance output matrix and cross-covariance matrix.

$$P_{y,k} = \sum_{i=0}^{2n} W_c^{(i)} [Y_{k|k-1}^i - \hat{y}_{k-1}][Y_{k|k-1}^i - \hat{y}_{k-1}]^T.$$

$$P_{xy,k} = \sum_{i=0}^{2n} W_c^{(i)} [\tilde{X}_{k|k-1}^i - \hat{x}_{k-1}][Y_{k|k-1}^i - \hat{y}_{k-1}]^T \text{ is the cross-covariance x-y.}$$

5.2 Compute the Kalman filter gain:

$$K_k = P_{xy,k} P_{y,k}^{-1}.$$

5.3 State estimate update:

$$\hat{x}_k = \hat{x}_{k|k-1} + K_k(y_k - \hat{y}_{k|k-1}).$$

5.4 State covariance matrix estimate update:

$$P_k = P_{x,k|k-1}^- - K_k P_{y,k} K_k^T.$$

[UAKF 6]. Correction measurement covariance matrices of noises:

6.1 Compute the output error:

$$\varepsilon_k = y_k - g(\hat{x}_k, u_k).$$

6.2 Compute the adjustment coefficient:

$$c_k = \frac{1}{L} \sum_{i=k-L+1}^{k} \varepsilon_k \varepsilon_k^T \text{ L is the window length (number of samples inside the window).}$$

6.3 Compute the covariance matrix of process noise:

$$q_{w,k} = K_k c_k K_k^T.$$

6.4 Compute the covariance matrix of the measurement noise:

$$r_{v,k} = c_k + \sum_{i=0}^{2n} W_c^{(i)} [Y_{k|k-1}^i - y_k + c_k][Y_{k|k-1}^i - y_k + c_k]^T.$$

For a better understanding of this algorithm, references [6–9] provide an excellent source of documentation.

The following two sets of tuned parameter values are used in MATLAB simulations for this algorithm:

- For the 3RC ECM Li-ion battery model: $\alpha = 0.05$; $\beta = 2$ (optimal value); $k = -1$; $L = 150$; $qw = I_{4 \times 4}$ (unity matrix); $rv = 0.1$; $Px = 10^{-10} I_{4 \times 4}$; SOCini = 70 (%); VarY = 0.001 (the variance of the noise level in the measurement output dataset used to test the robustness); $\eta = 0.78$ for charging cycle; and $\eta = 0.9$ for discharging cycle.
- For the Li-ion battery Simscape model: $\alpha = 1$; $\beta = 2$ (optimal value); $k = 0$; $L = 300$; $qw = 0.0001$; $rv = 0.00019$; $Px = 10^{-10}$; SOCini = 70 (%); VarY = 0.001; $\eta = 0.765$ for charging cycle; and $\eta = 0.865$ for discharging cycle.

3.2.1. MATLAB Implementation and Simulation Results for 3RC ECM Battery Model-AUKF SOC Estimator Accuracy and Robustness Scenarios

- Scenario R0. The MATLAB simulation results are shown in Appendix A.1, Figure A18a–c, and the statistical criteria values are given in Table 1.

  Performance analysis:

  ○ SOC accuracy is great and battery output voltage prediction is excellent.
  ○ The residual error is quite close to 0.6%, which is an excellent result.

- Scenario R1. The MATLAB simulations result are shown in Appendix A.1, Figure A19a,b, and the statistical criteria values are given in Appendix A.2, Table A1.

    Performance analysis:

    ◯   SOC of high accuracy.
    ◯   The residual error is quite close to 0.5%, which is an excellent result.

- Scenario R2. The MATLAB simulation results are shown in Figure 6, and statistical criteria values are given in Appendix A.2, Table A2.

    Performance analysis:

    ◯   SOC accuracy is good.
    ◯   The residual error is quite close to 5%, which is a weak performance.

- Scenario R3. The MATLAB simulation results are shown in Appendix A.1, Figure A20a–c, and the statistical criteria values are given in Appendix A.2, Table A3.

    Performance analysis:

    ◯   SOC accuracy is great.
    ◯   The residual error is quite close to 0.48%, which is excellent.

- Scenario R4. The MATLAB simulations result for fourth scenario is depicted in Figure 7a,b and the statistical criteria values are given in Appendix A.2, Table A4.

    Performance analysis:

    ◯   SOC accuracy is great.
    ◯   The residual error is quite close to zero in steady state, so an excellent result.

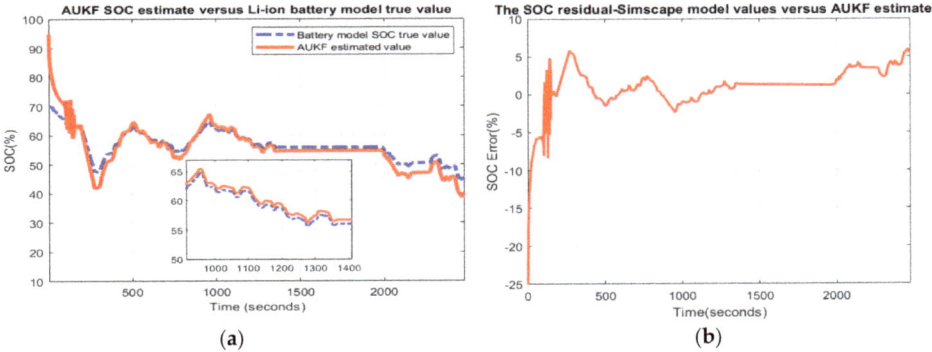

Figure 6. Robustness to simultaneous changes, SOCini = 1, Qnom = 4.2 Ah (ageing effects); (**a**) AUKF SOC value versus battery model true value; (**b**) SOC residual.

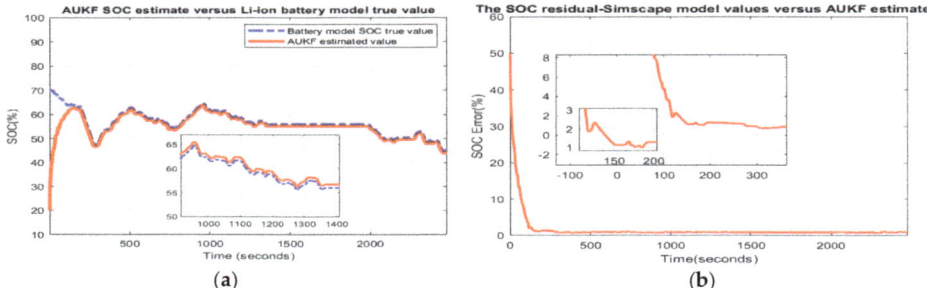

Figure 7. Robustness to simultaneous changes, SOCini = 0.2, and output temperature profile changes; (**a**) AUKF SOC value versus battery model true value; (**b**) SOC residual.

3.2.2. MATLAB Implementation and Simulation Results for Simulink Simscape Battery Model—AUKF SOC Estimator Accuracy and Robustness Scenarios

- Scenario R0. The MATLAB simulation results for this scenario are shown in Appendix A.1, Figure A21a–c, and the statistical criteria values are given in Table 1.

  Performance analysis:

  ○ SOC accuracy is great and battery output voltage prediction is excellent.
  ○ The residual error is quite close to 2%, which is a great result.

- Scenario R1. The MATLAB simulation results for the first scenario are depicted in Appendix A.1, Figure A22a–c, and the statistical criteria values are given in Appendix A.2, Table A1.

  Performance analysis:

  ○ SOC accuracy is great.
  ○ The residual error is quite close to 2%, which is an excellent result.

- Scenario R2. The MATLAB simulation results for the second scenario are visible in Figure 8 and statistical criteria values are given in Appendix A.2, Table A2.

  Performance analysis:

  ○ SOC accuracy is good.
  ○ The residual error is quite close to 7%, which is a bad result.

- Scenario R3. The MATLAB simulation results for third scenario are shown in Appendix A.1, Figure A23a–c, and the statistical criteria values are given in Appendix A.2, Table A3.

  Performance analysis:

  ○ SOC accuracy is great.
  ○ The residual error is quite close to 1%, which is excellent.

- Scenario R4. The MATLAB simulation results for the fourth scenario are depicted in Figure 9 and the statistical criteria values are given in Appendix A.2, Table A4.

  Performance analysis:

  ○ SOC accuracy is great.
  ○ The residual error is quite close to 1.2%, so is excellent.

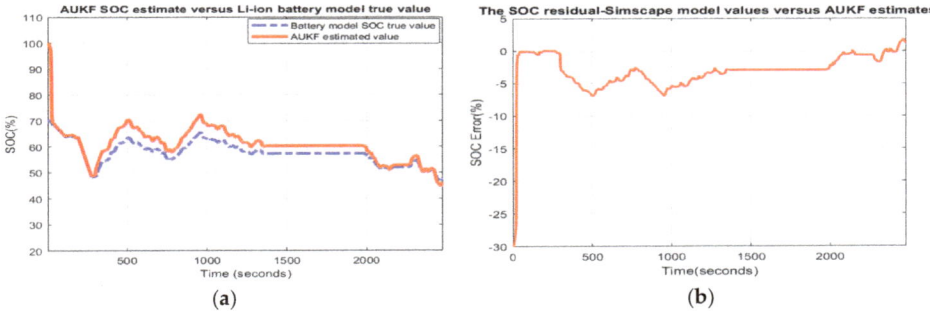

Figure 8. Robustness to simultaneous changes, SOCini = 1, Qnom = 4.2 Ah (ageing effects); (**a**) AUKF SOC value versus battery model true value; (**b**) SOC residual.

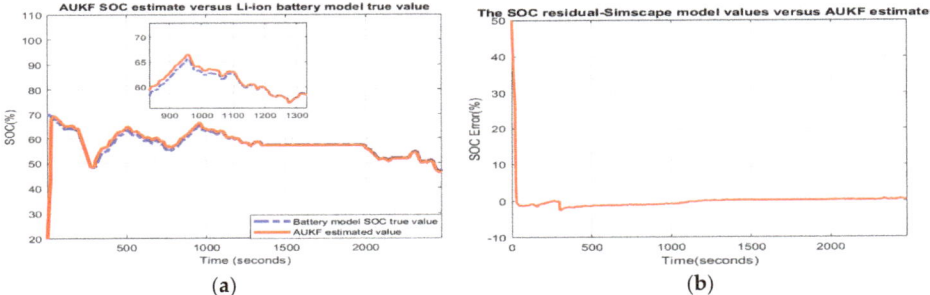

Figure 9. Robustness to simultaneous changes, SOCini = 0.2, and output temperature profile changes; (**a**) AUKF SOC value versus battery model true value; (**b**) SOC residual.

In general, based on the MATLAB simulation results of AUKF SOC performance obtained for each model, it is relevant that the AUKF SOC estimator works well in four scenarios. The simulations from the second scenario (R2) reveal that AUKF is more sensitive to the effects of ageing in both models, so it is difficult to distinguish at this stage. The values of the statistical criteria can provide sufficient information to make a correct delimitation.

3.3. Particle Filter SOC Estimator

In this section, a second nonlinear SOC estimator, namely PFE SOC, is chosen to achieve a high precision SOC for both models of Li-ion batteries, which makes possible a complete, relevant and credible analysis of the performance of all three SOC estimators proposed in our research. It is used "to estimate the states, which approximate the probability density function of a non-linear, non-Gaussian system by using the Monte-Carlo simulation technique", as is mentioned in [11].

3.4. PFE SOC Brief Presentation

There is a substantial similarity between the non-linear estimator PFE SOC [10,11] and the first two SOC estimators presented in the previous subsections, i.e., AEKF SOC and AUKF SOC, due to the same "prediction-corrector" structure identified in all three. Therefore, it is easy to anticipate that the PFE SOC estimator updates in a "recursively way" the state estimate and then finds the innovations driving a stochastic process based on a sequence of observations (measurement output dataset), as is shown in detail in the original work [11]. In [11] it is stated that the PFE SOC estimator accomplishes this objective by "a sequential Monte Carlo method (bootstrap filtering), a technique for implementing a recursive Bayesian filter by Monte Carlo simulations", which is also mentioned in [4]. After the initialization stage of the algorithm, in the second stage (i.e., "the prediction phase"), the state estimates of the process are used to predict and to "smooth" the stochastic process. As a result of the prediction, innovations are useful for estimating the parameters of the linear or nonlinear dynamic model [4,11]. The basic idea disclosed in [4] is that any probability distribution function (pdf) of a random state variable x can be approximated by a set of samples (particles), similar to what sigma points do in the AUKF SOC estimator developed in Section 3.2. Each particle has one set of values for each process state variable x. The novelty of the PFE SOC estimator is its ability to represent any arbitrary pdf, even if for non-Gaussian or multi-modal pdfs [4,11]. In conclusion, the nonlinear design of the SOC PFE estimator has a similar approach to that of the AUKF SOC design, as long as a local linearization technique is not required, as in the case of AEKF SOC, or "any raw functional approximation" [4,11]. Furthermore, the PFE SOC "can adjust the number of particles to match available computational resources, so a trade-off between accuracy of estimate and required computation" [11]. Moreover, it is "computationally compliant even with complex, non-linear, non-Gaussian models, as a trade-off between approximate solutions to complex nonlinear dynamic model versus exact solution to approximate dynamic model" [11]. To get a better insight into this estimation technique, the original paper [11] can be accessed. Since, the current research work follows the same PFE design procedure steps as in [11], our focus is directed only at the implementation aspects.

3.5. PFE SOC Parameters' Setup

The following two sets of tuned parameter values are used in MATLAB simulations for this algorithm:

- For the 3RC ECM Li-ion battery model: Np = 1000 (total number of particles); qw = $10^{-6} I_{4\times 4}$ ($I_{4\times 4}$ is a 4×4 identity matrix) in scientific notation (process state variables noise covariance matrix); rv = 0.0001 (measurement output noise); SOCini = 70 (%); VarY = 0.001 (variance of the noise level in the measurement output dataset used to test the robustness); VarX1 = VarX2 = VarX3 = VarX4 = 0.01 (variance in the initial values of the states variables); η = 0.8 for charging cycle; and η = 0.82 for discharging cycle.
- For the Li-ion battery Simscape model: Np = 1000 (total number of particles); qw = 10^{-7} for SOC covariance noise; rv = 0.0001 (measurement output noise level); SOCini = 70 (%); VarY = 0.001 (the variance of the noise level in the measurement output dataset used to test the robustness); VarX1 = 0.004 (variance in SOCini); η = 0.76 for charging cycle; and η = 0.78 for discharging cycle.

3.6. MATLAB Simulation Results for 3RC ECM Battery Model—PFE SOC Estimator Accuracy and Robustness Scenarios

- Scenario R0. The MATLAB simulation results for this scenario are shown in Appendix A.1, Figure A24a–c, and the statistical criteria values are given in Table 1.

    Performance analysis:

    ○ SOC accuracy is good and battery output voltage prediction is good.
    ○ The residual error is quite close to 8%, which is weak.

- Scenario R1. The MATLAB simulation results for the first scenario are shown in Appendix A.1, Figure A25a–c, and the statistical criteria values are given in Appendix A.2, Table A1.

    Performance analysis:

    ○ SOC accuracy is good.
    ○ The residual error is quite close to 10%, which is weak.

- Scenario R2. The MATLAB simulation results for the second scenario are visible in Figure 10 and statistical criteria values are given in Appendix A.2, Table A2.

Performance analysis:

- ○ SOC accuracy is weak.
- ○ The residual error is quite close to 10%, which is bad.

- Scenario R3. The MATLAB simulation results for the third scenario are revealed in Appendix A.1, Figure A26a–c, and the statistical criteria values are given in Appendix A.2, Table A3.

Performance analysis:

- ○ SOC accuracy is good.
- ○ The residual error is quite close to 4%, which is weak.

- Scenario R4. The MATLAB simulation results for fourth scenario are depicted in Figure 11 and the statistical criteria values are given in Appendix A.2, Table A4.

Performance analysis:

- ○ SOC accuracy is weak.
- ○ The residual error is quite close to 20%, which is bad.

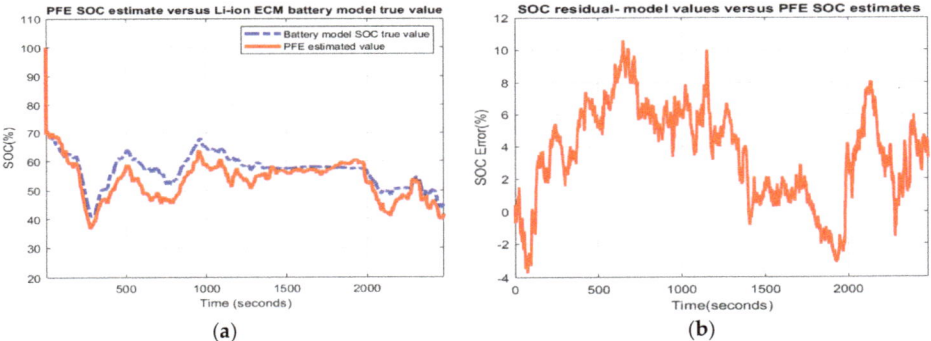

Figure 10. Robustness to simultaneous changes, SOCini = 1, Qnom = 4.2 Ah (ageing effects); (**a**) PFE SOC value versus 3RC ECM battery model true value; (**b**) SOC residual.

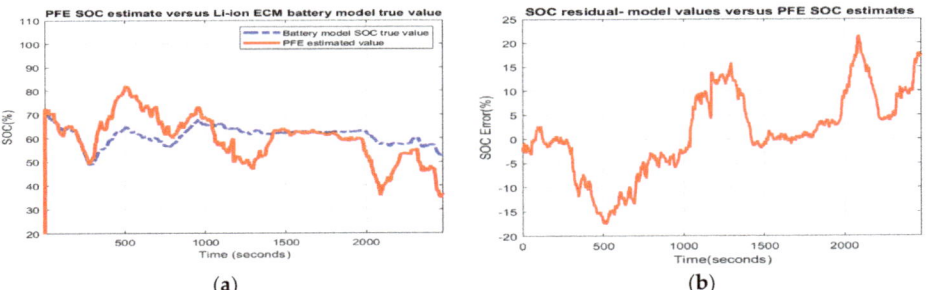

Figure 11. Robustness to simultaneous changes, SOCini = 0.2, and output temperature profile changes; (**a**) PFE SOC value versus 3RC ECM Li-ion battery model true value for changes only in Rin; (**b**) SOC residual for 20% changes in Rin.

3.7. MATLAB Simulation Results for Simulink Simscape Battery Model—PFE SOC Estimator Accuracy and Robustness Scenarios

- Scenario R0. The MATLAB simulation results for this scenario are shown in Appendix A.1, Figure A27a–c, and the statistical criteria values are given in Table 1.

  Performance analysis:

  ○ SOC accuracy is good.
  ○ The residual error is quite close to 4%, which is weak.

- Scenario R1. The MATLAB simulation results for the first scenario are revealed in Appendix A.1, Figure A28a–c, and the statistical criteria values are given in Appendix A.2, Table A1.

  Performance analysis:

  ○ SOC accuracy is good.
  ○ The residual error is quite close to 2%, which is good.

- Scenario R2. The MATLAB simulation results for the second scenario are depicted in Figure 12 and statistical criteria values are given in Appendix A.2, Table A2.

  Performance analysis:

  ○ SOC accuracy is great.
  ○ The residual error is quite close to 6%, which is weak.

- Scenario R3. The MATLAB simulation results for the third scenario are visible in Appendix A.1, Figure A29a–c, and the statistical criteria values are given in Appendix A.2, Table A3.

  Performance analysis:

  ○ SOC accuracy is great.
  ○ The residual error is quite close to 3%, which is good.

- Scenario R4. The MATLAB simulation results for the fourth scenario are shown in Figure 13a,b for changes in Kp, and Figure 13c,d for changes in Rin. The statistical criteria values are given in Appendix A.2, Table A4.

  Performance analysis:

  ○ SOC accuracy is weak for changes in Rin and good for changes in Kp.
  ○ The residual error is quite close to 12% for 10% changes in Rin, which is bad.

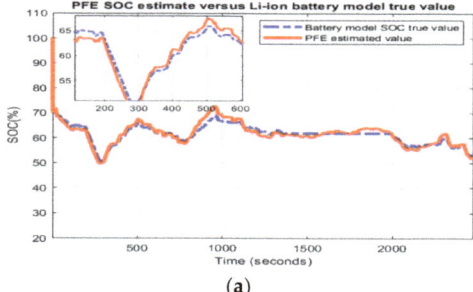 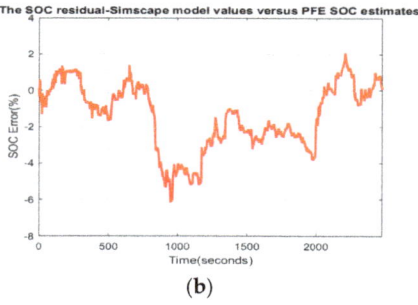

(a)   (b)

Figure 12. Robustness to simultaneous changes, SOCini = 1, Qnom = 4.2 Ah (ageing effects); (**a**) PFE SOC value versus battery model true value; (**b**) SOC residual.

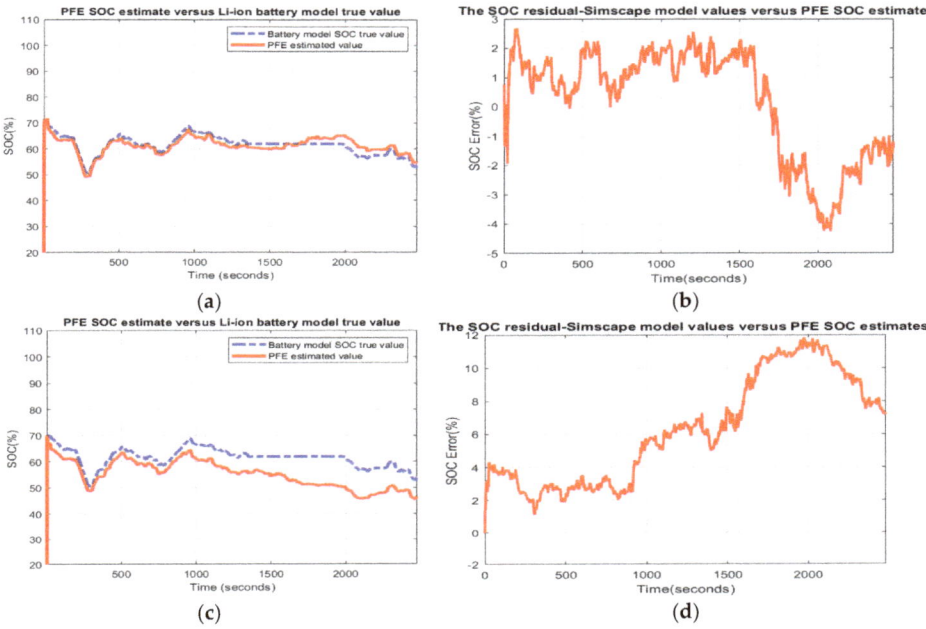

Figure 13. Robustness to simultaneous changes, SOCini = 0.2, and output temperature profile changes; (**a**) PFE SOC value versus battery model true value for changes only in Kp; (**b**) SOC residual for changes only in Kp; (**c**) PFE SOC value versus battery model true value for 10% changes only in Rin; (**d**) SOC residual for changes only in Rin.

Similarly, for the first and second SOC estimators, the MATLAB simulation results of the PFE SOC performance obtained for each model reveal that the PFE SOC estimator works satisfactorily in four scenarios (R0, R1, R2, R3) for the Simscape model, and three scenarios (R0, R1, R3) for the 3RC EMC model. Thus, it is confirmed again that the Simulink model is suitable for use as a support for designing and implementing in a real-time MATLAB environment of SOC estimators in HEV applications.

4. Discussion

This research work has been beneficial for us, as our research experience was considerably improved, and we learned some useful lessons for the future. The preliminary results obtained so far in the design, modelling, implementation and validation of Li-ion batteries, development and implementation of real-time SOC estimation algorithms are enriched continuously and supplemented based on a new experience and a considerable routine in using one of the most powerful pieces of software for real-time design and implementation, such as MATLAB and Simulink. In the following are summarized some of the most relevant aspects that have attracted our attention and enriched our research experience in the field so far. In the next five subsections, the performance analysis of each SOC estimator, in terms of SOC convergence speed, real-time implementation, accuracy and robustness performance analysis, is based on the data shown in Tables 1 and 2 and Figures A12, A15, A18 and A21 from Appendix A.1, corresponding to the scenario R0, for an SOC initial value, SOCini = 70%.

Table 2. The response speed convergence of AEKF SOC, UAKF SOC and PFE SOC (in seconds). Qnom = 4.2 Ah (30% degradation of battery due to ageing effects) VarY = 0.001.

Li-Ion Battery Model	3RC ECM			Simscape		
SOC Estimator	SOCini = 0.2	SOCini = 0.4	SOCini = 0.9	SOCini = 0.2	SOCini = 0.4	SOCini = 0.9
AEKF	188	10	21	25	23	25
AUKF	194	140	170	36	34	30
PFE	23	22	35	32	26	28

4.1. SOC Estimators' Convergence Speed

The analysis of Li-ion battery response convergence speed for all three SOC estimators and each model can be done visually by examining the graphs related to SOC or based on the benchmark represented in Table 2. The data from Table 2 correspond to the worst-case scenario that could happen in "real life", since they analyse a Li-ion capacity degradation by 30% due to ageing effects and for changes in the "guessed" value of initial SOC. Moreover, some noise in measurement output data (battery terminal voltage) has a variance VarY = 0.001. From data provided in Table 2, it can be said that for the 3RC ECM Li-ion battery model SOC the PFE is much faster compared to the other two estimators, followed by the AEKF SOC estimator. For the Li-ion battery Simscape model the AEKF SOC estimator is faster than the other two competitors, followed quite closely by the PFE SOC estimator. By a rigorous analysis of data collected in Table 1, it can conclude that all three SOC estimators perform better for the Li-ion battery Simscape model; the response convergence speed is faster than for the 3RC ECM Li-ion battery model.

4.2. SOC Estimation Accuracy

A rigorous analysis of SOC estimation accuracy performance can be performed using the information extracted from the SOC residual corresponding to the first scenario, Ro, i.e., for a SOCini value of 0.7 and all other parameters of Li-ion battery adjusted to the nominal values, as shown in Table 3. Moreover, the SOC accuracy is strongly related to the battery model accuracy. Since both Li-ion battery models are exactly accurate, as was shown in Part 1, an excellent efficiency for all three estimators based on both battery models can be anticipated. The second assessment procedure of SOC estimation accuracy of each SOC estimator can also be carried out based on all six statistical criteria values obtained from Table 1. Moreover, a complete performance analysis consists of analysing the information provided by each SOC residual value and using statistical criteria. By inspecting the statistical criteria values, column by column, for each model, the AEKF SOC estimator based on the Simulink model behaves slightly better than two other competitors, followed by SOC estimators PFE and AUKF.

Table 3. The Li-ion SOC estimator accuracy based on the SOC residual error (%).

Li-Ion Battery Model	3RC ECM			Simscape		
SOC Estimator	AEKF	AUKF	PFE	AEKF	AUKF	PFE
Figure A11	<1.6					
Figure A14				<0.4		
Figure A17		<0.8				
Figure A20					<2	
Figure A23			<8			
Figure A26						<2
Result	Good	High	Low	High	Good	Good

On the other hand, for a 3RC ECM battery model the AUKF behaves better, followed closely by AEKF and PFE. By far, combining the results obtained in Tables 1 and 3, it can be said that the AEKF

SOC estimator has better performance for the Simscape battery model, followed quite closely by the AUKF SOC estimator. For a 3RC ECM battery model it is the AUKF that performs better, followed by AEKF and PFE SOC estimators. However, since the values of statistical criteria extracted from Table 1 are close to each other for most of them, it is difficult now to make a net difference between the performance of all three SOC estimators. Moreover, sometimes it is difficult in some situations to make an interpretation that is approximative of each statistical criteria value. Still, in some cases, due to unsuitable values for the tuning parameters, the AEKF, AUKF and PFE SOC estimates are biased. Regarding all three SOC estimators, we observed that the SOC accuracy depends on a "trial and error" empirical adjustment procedure of tuning parameter values. Unfortunately, this procedure takes much time. Moreover, a new readjustment procedure is required when changing the driving conditions and SOC initial value, as well as when ageing and temperature effects take place. The adopted versions of AEKF and AUKF, due to their adaptive features, attenuate the tuning procedure of the parameters significantly.

4.3. SOC Estimator—Measurement Noise Filtration

A critical aspect observed in this research is the measurement noise filtration by all three estimators. Only the AEKF and AUKF have this ability to filtrate the measurement noise due to the noise correction step in each algorithm, compared to the PFE SOC estimator that does not have this feature.

4.4. SOC Estimators—Real Time Implementation

As was mentioned in the previous section, due to their predictor–corrector structure, each SOC estimator becomes a recursive algorithm, more straightforward to implement in real-time and very efficient in terms of computation. Both Li-ion battery models are also simple, easy to design and quickly deploy, especially the Simscape battery model based directly on the manufacturer's battery specifications. Besides, MATLAB-Simulink software platform provides a valuable and practical Simscape/SimPower Systems library, helpful for use in design and implementation of different HEV and EV powertrain configurations.

4.5. SOC Estimator Robustness Performance Analysis—Statistical Criteria

The values of statistical criteria from Table 1 provide the SOC accuracy of both battery models concerning ADVISOR estimate, beneficial for Li-ion battery model validation performed for an FTP-75 driving cycle profile test. The statistical criteria values from Tables A1–A4 are valuable for analysing the SOC robustness performance of all three SOC estimators. Based on the information extracted from Tables A1–A4 for each SOC estimator, it seems that AEKF SOC is more robust compared to the other two SOC estimators, as is quite evident for the Simulink model. Unfortunately, it is diffficult to make a complete performance analysis by comparison of the results obtained by similar SOC estimators reported in the literature. This happens since many researchers use different input current profiles and various statistical criteria that do not match with those used in our research. However, for the cases that match with our driving cycle profile test, the information collected in Tables 1 and A1, Tables A2–A4 can be useful for analysing all similar situations. Thus, the present research work can be a valuable source of inspiration for readers and researchers.

5. Conclusions

In the current research paper, the following most relevant contributions of the authors can be highlighted:

- Adaptive Extended Kalman Filter SOC estimator with fading feature and covariance matrices of noises correction—brief presentation and MATLAB application.
- Adaptive Unscented Kalman Filter SOC estimator with covariance matrices of process and measurement noise correction—design and MATLAB implementation.

- Adaptive Particle Filter SOC estimator—brief presentation and MATLAB application.
- MATLAB SOC simulations for all three SOC estimators.
- Performance analysis for five scenarios (SOC accuracy and robustness)—Tables 1 and A1, Tables A2–A4 for six statistical errors defined in Part 1 [20], namely RMSE, MSE, MAE, std, MAPE and R-squared.

Based on six statistical criteria values for all three SOC estimators, as a behavior response to an FTP-75 driving cycle profile test, it was possible to choose, from all three competitors, the most suitable SOC estimator. The result of the overall performance analysis indicates that the AEKF SOC estimator performs better than the other two competing SOC estimators.

In future work, our investigations will continue to improve the design and implementation approach by using fuzzy logic, neural networks and learning machine methods from artificial intelligence field.

Author Contributions: R.-E.T. has contributed for, algorithm conceptualization, software, original draft preparation and writing it. N.T. has contributed for battery models investigation and validation, performed MATLAB simulations and formal analysis of the results. M.Z. has contributed for project administration, supervision, and results visualization. S.-M.R. has contributed for methodology, data curation and supervision. All authors have read and agreed to the published version of the manuscript.

Funding: This research received no external funding.

Acknowledgments: In Research funding (discovery grant) for this project from the Natural Sciences and Engineering Research Council of Canada (NSERC) is gratefully acknowledged.

Conflicts of Interest: The authors declare no conflict of interest.

Abbreviations

Ni-Cad	nickel cadmium
Ni-MH	nickel metal hydride
Li-ion Co	lithium-ion cobalt
EV	electric vehicle
HEV	hybrid electric vehicle
BMS	battery management system
ADVISOR	advanced vehicle simulator
EPA	environmental protection agency
UDDS	urban dynamometer driving schedule
FTP-75	Federal test procedure at 75 [degrees F]
SMO	sliding mode observer
LOE	linear observer estimator
RMSE	root mean squared error
MSE	mean squared error
MAE	mean absolute error
MAPE	mean absolute percentage error
R^2/R-squared	coefficient of determination
std (σ)	standard deviation
OCV	open-circuit voltage
SOC	state of charge
SOE	state of energy
SOH	state of health
DOD	depth of discharge
NREL	National Renewable Energy Laboratory
EKF	extended Kalman filter
AEKF	adaptive extended Kalman filter
UKF	unscented Kalman filter
AUKF	adaptive unscented Kalman filter
PFE	particle filter estimator

Appendix A

Appendix A.1. Figures

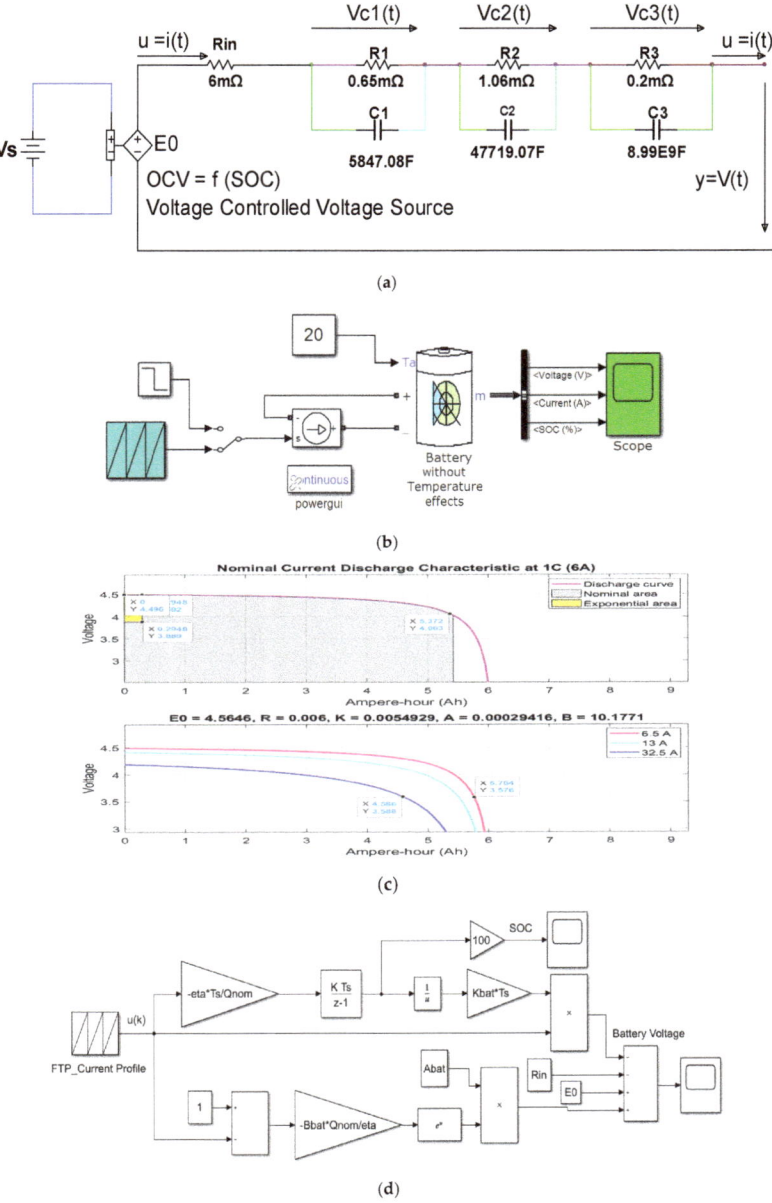

Figure A1. The Li-ion Battery—3RC ECM and Simscape models: (**a**) Li-ion battery represented in NI Multisim 14.1 editor (see Figure 6—Part 1); (**b**) Simscape Simulink diagram of Li-ion battery (see Figure A19—Part 1); (**c**) Simscape SAFT Li-ion battery nominal current discharge characteristic @1C (6A) (top side view); @6.5A, 13A and 32.5A. (**d**) Simulink diagram of Li-ion Simscape generic model.

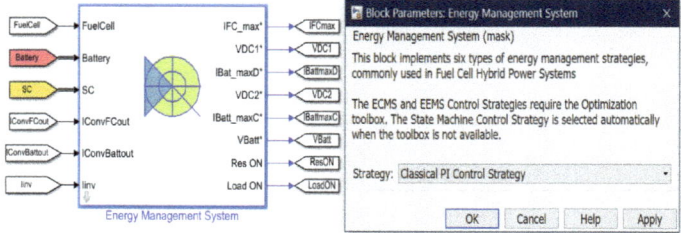

Figure A2. EMS—Classical PI Control Strategy setup.

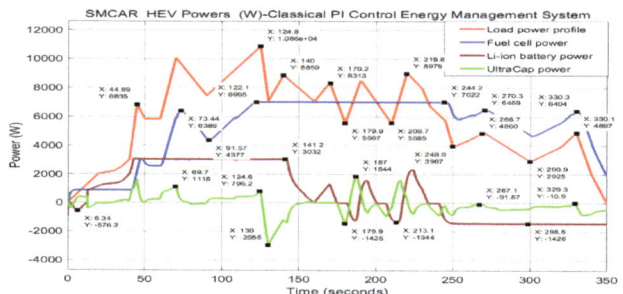

Figure A3. SMCAR HEV Powers—Classical PI Control EMS.

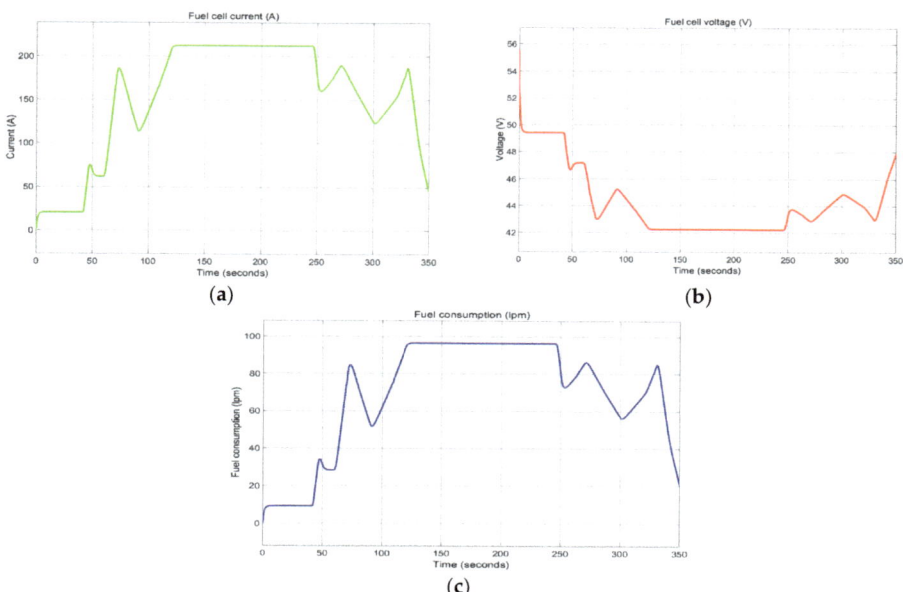

Figure A4. Fuel cell. (**a**) FC current; (**b**) FC voltage; (**c**) FC consumption.

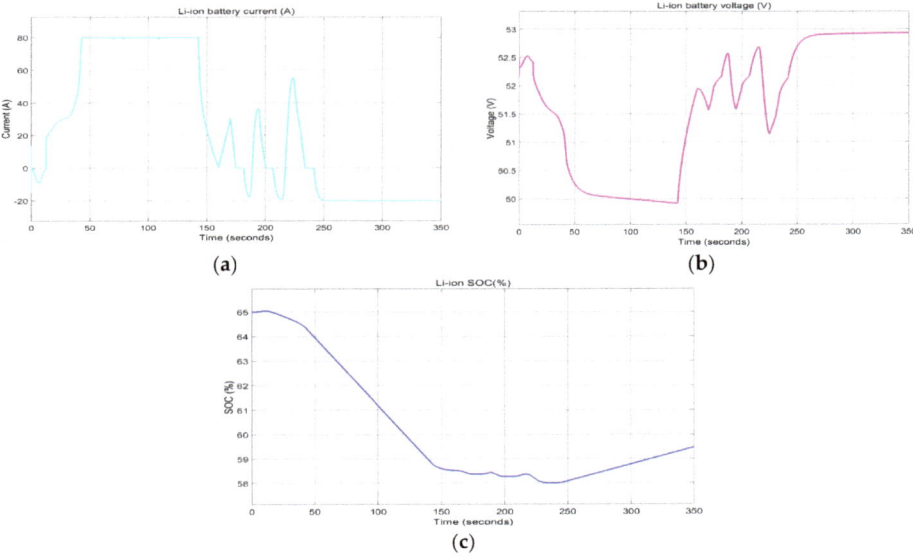

Figure A5. Li-ion battery specific variables. (**a**) Battery current; (**b**) Battery voltage; (**c**) Battery SOC.

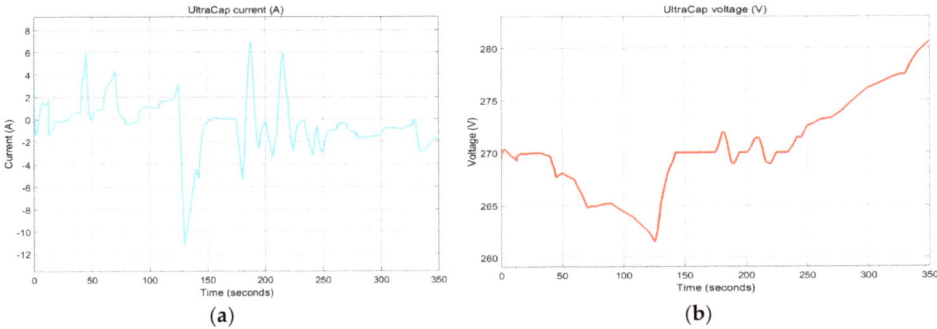

Figure A6. SC specific variables. (**a**) SC current variation; (**b**) SC voltage.

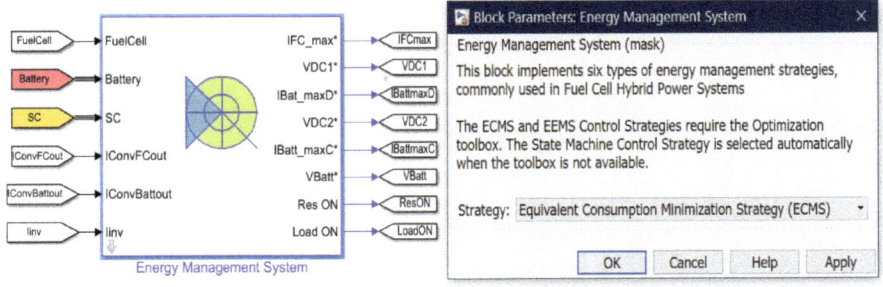

Figure A7. EMS—Equivalent Consumption Minimization Strategy setup.

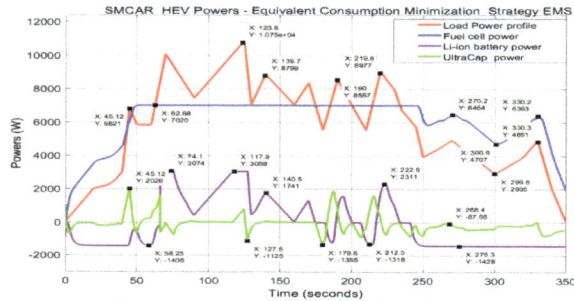

Figure A8. SMCAR HEV Powers—Equivalent Consumption Minimization EMS.

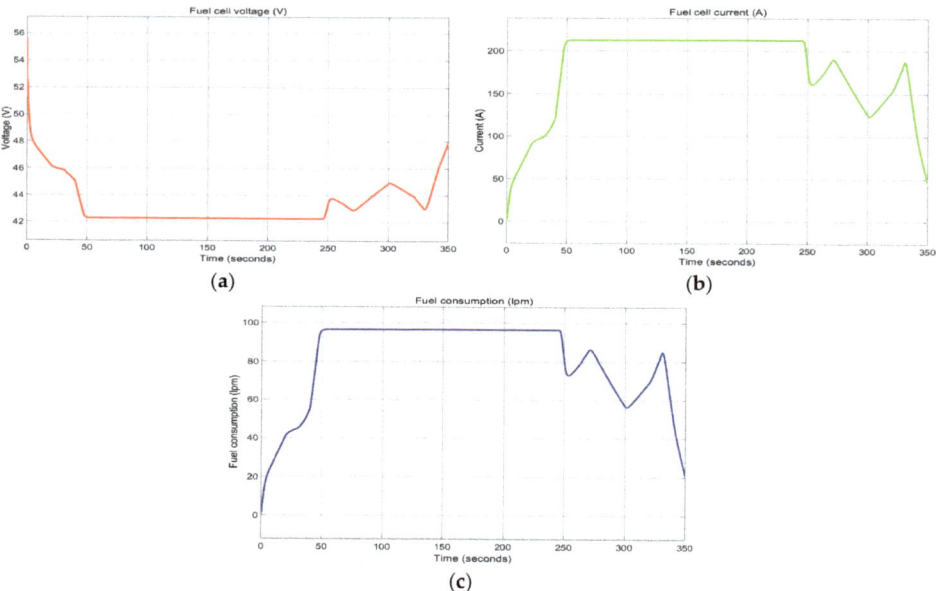

Figure A9. Fuel cell. (**a**) FC current; (**b**) FC voltage; (**c**) FC consumption.

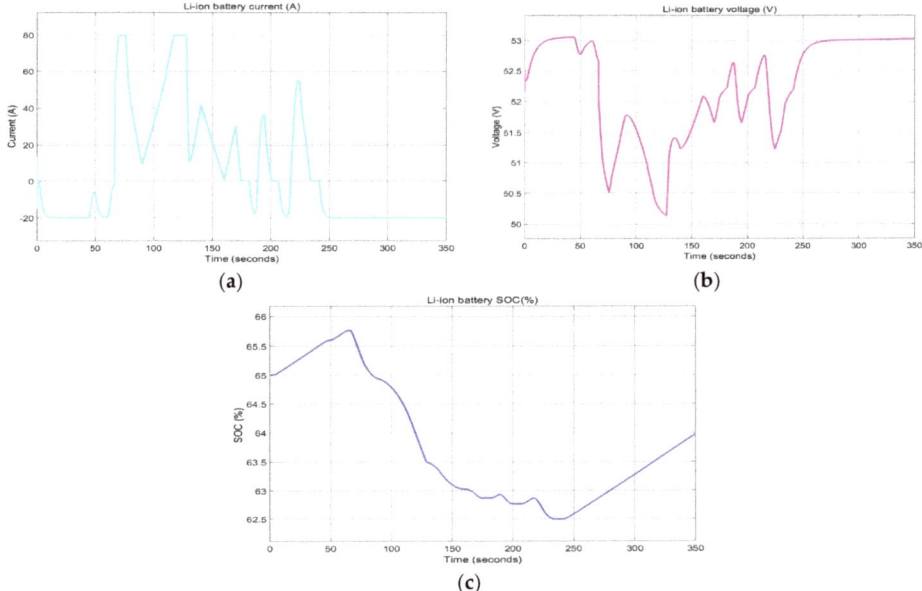

Figure A10. Li-ion battery specific variables. (**a**) Battery current; (**b**) Battery voltage; (**c**) Battery SOC.

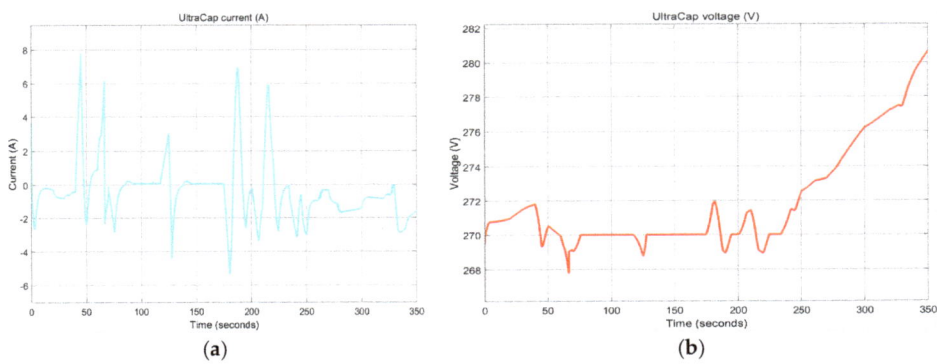

Figure A11. SC specific variables. (**a**) SC current variation; (**b**) SC voltage.

- 3RC EMC Li-Ion Battery Model—A EKF SOC Estimator.

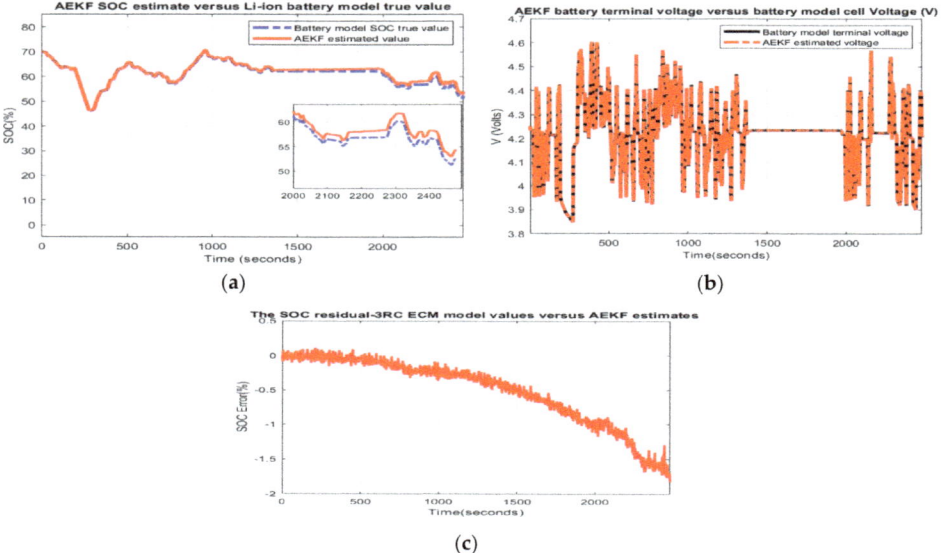

Figure A12. The MATLAB simulation results for Simulink Simscape Li-ion battery. (**a**) The AEKF SOC value versus battery model SOC true value; (**b**) The AEKF terminal output voltage versus battery model terminal output voltage true value; (**c**) The SOC residual.

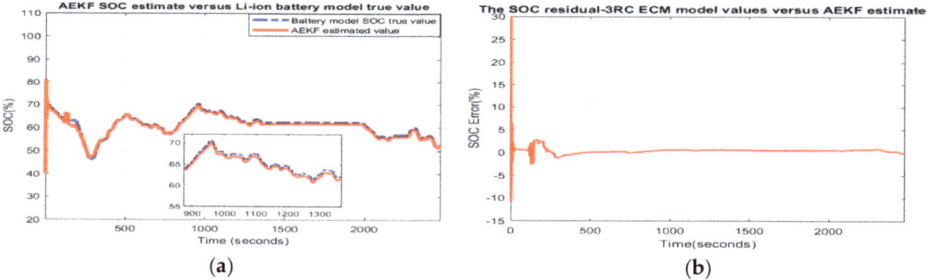

Figure A13. Robustness to changes in SOC initial value—SOCini = 0.4. (**a**) AEKF SOC estimate versus battery SOC true value; (**b**) SOC residual.

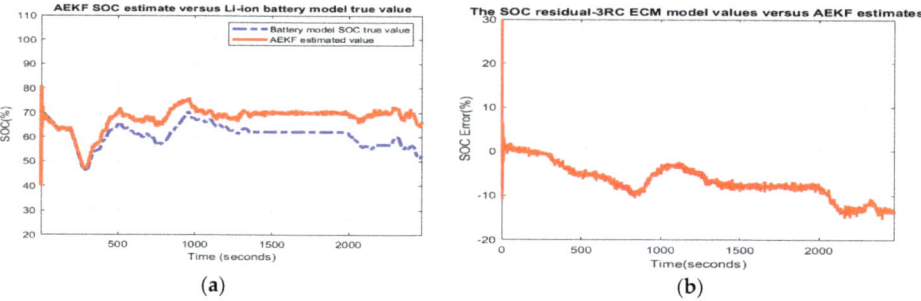

Figure A14. Robustness to simultaneous changes, SOCini = 0.4, σ noise = 0.01; (**a**) AEKF SOC value versus battery model true value; (**b**) SOC residual.

- 3 RC AEKF Li-ion Battery Simscape Model.

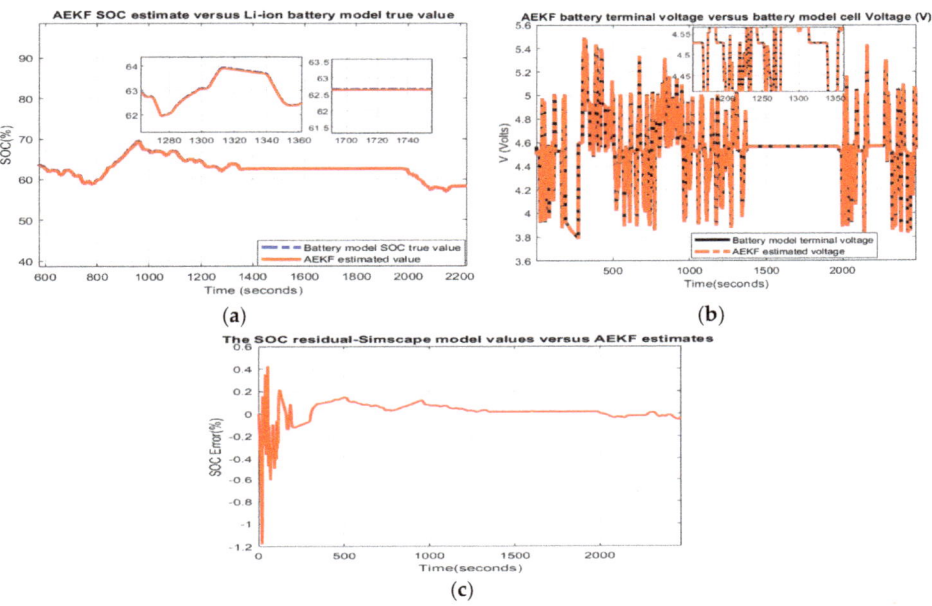

Figure A15. The MATLAB simulation results for Simulink Simscape Li-ion battery. (**a**) The AEKF SOC value versus battery model SOC true value; (**b**) The AEKF terminal output voltage versus battery model terminal output voltage true value; (**c**) The SOC residual.

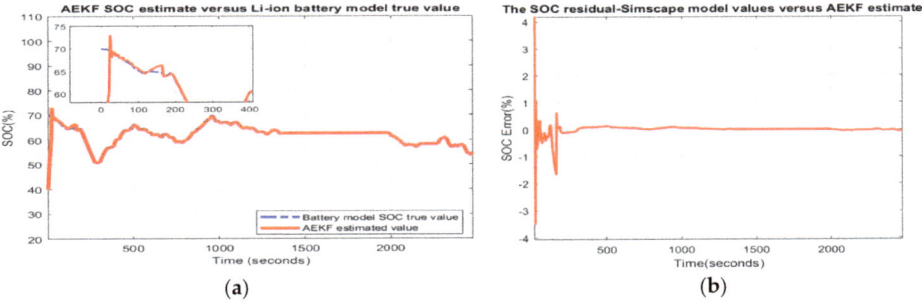

Figure A16. Robustness to changes in SOC initial value—SOCini = 0.4. (**a**) AEKF SOC estimate versus battery SOC true value; (**b**) SOC residual.

Batteries **2020**, *6*, 41

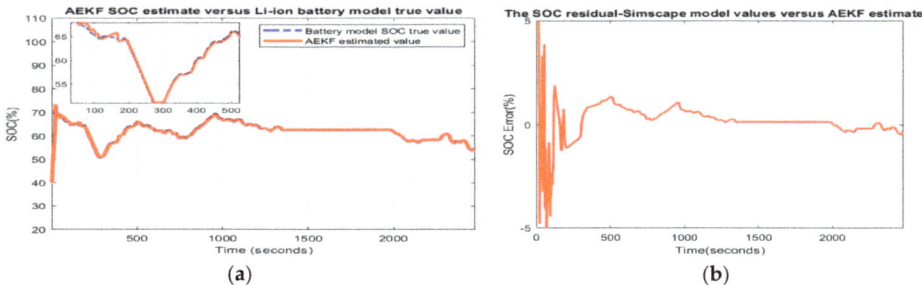

(a)   (b)

Figure A17. Robustness to simultaneous changes, SOCini = 0.4, σ noise = 0.01; (**a**) AEKF SOC value versus battery model true value; (**b**) SOC residual.

- AUKF 3RC EMC Li-ion Battery Model

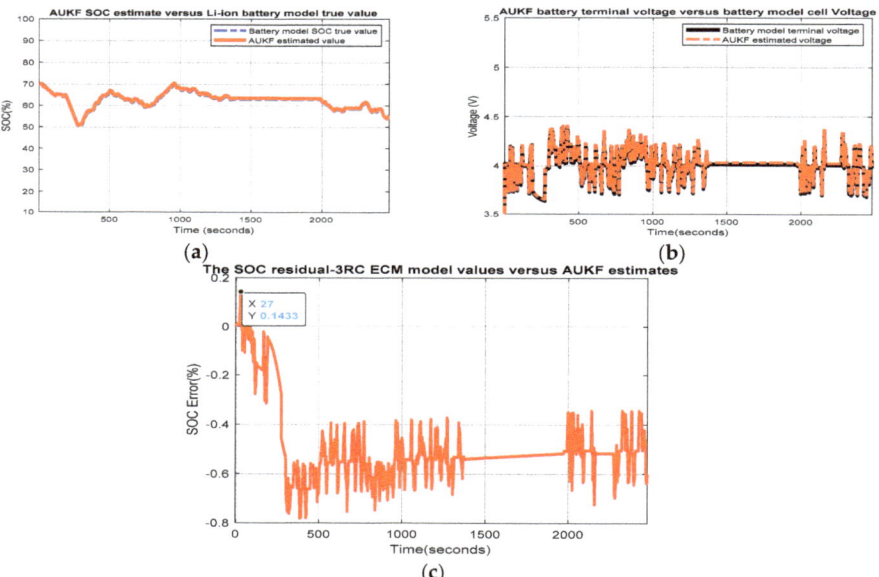

Figure A18. The MATLAB simulation results for Simulink 3RC ECM Li-ion battery. (**a**) The AUKF SOC value versus battery model SOC true value; (**b**) The AUKF terminal output voltage versus battery model terminal output voltage true value; (**c**) The SOC residual.

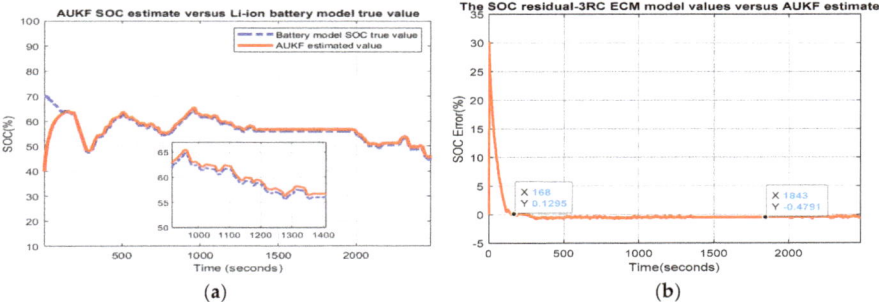

Figure A19. Robustness to changes in SOC initial value—SOCini = 0.4. (**a**) AUKF SOC estimate versus battery SOC true value; (**b**) SOC residual.

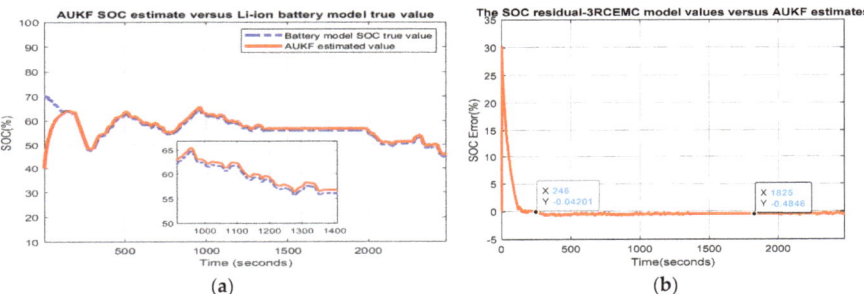

Figure A20. Robustness to simultaneous changes, SOCini = 0.4, σ noise = 0.01; (**a**) AUKF SOC value versus battery model true value; (**b**) SOC residual.

- AUKF Li-ion Battery Simscape Model

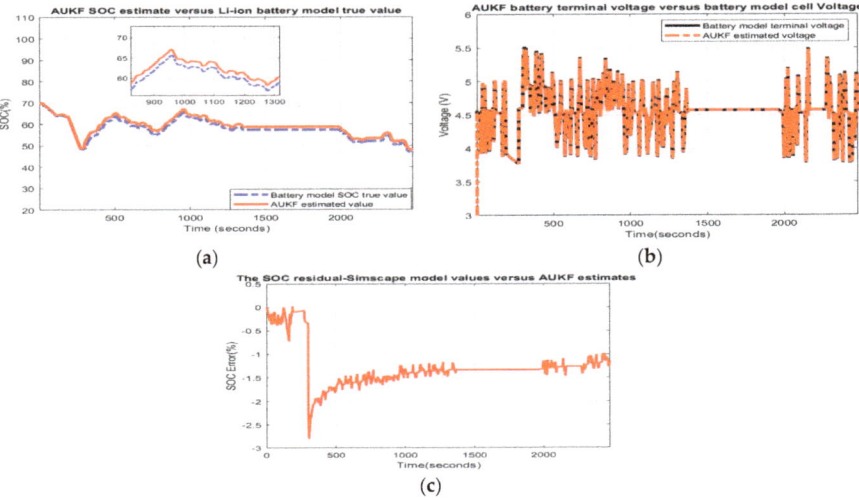

Figure A21. The MATLAB simulation results for Simulink Simscape Li-ion battery. (**a**) The AUKF SOC value versus battery model SOC true value; (**b**) The AUKF terminal output voltage versus battery model terminal output voltage true value; (**c**) The SOC residual.

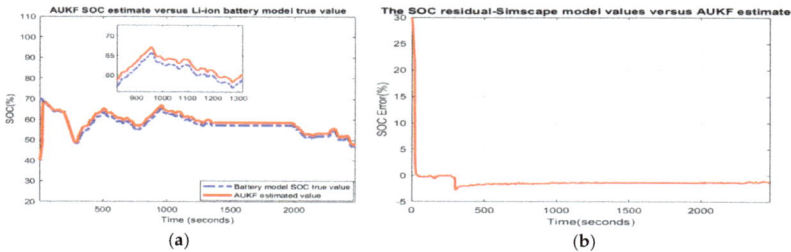

Figure A22. Robustness to changes in SOC initial value—SOCini = 0.4. (**a**) AUKF SOC estimate versus battery SOC true value; (**b**) SOC residual.

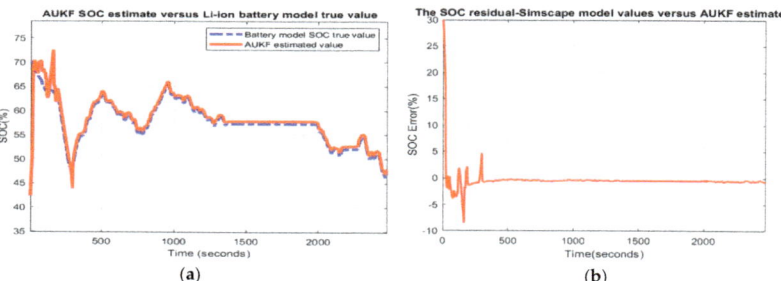

Figure A23. Robustness to simultaneous changes, SOCini = 0.4, σ noise = 0.01; (**a**) AUKF SOC value versus battery model true value; (**b**) SOC residual.

- PFE 3RC ECM Li-ion Battery Model

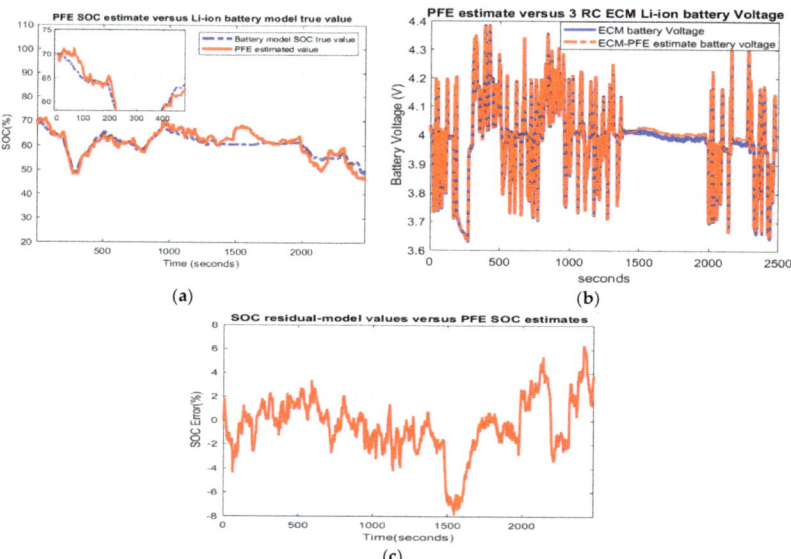

Figure A24. The MATLAB simulation results for Simulink Simscape Li-ion battery. (**a**) The PFE SOC value versus battery model SOC true value; (**b**) The PFE terminal output voltage versus battery 3RC ECM model terminal output voltage true value; (**c**) The SOC residual.

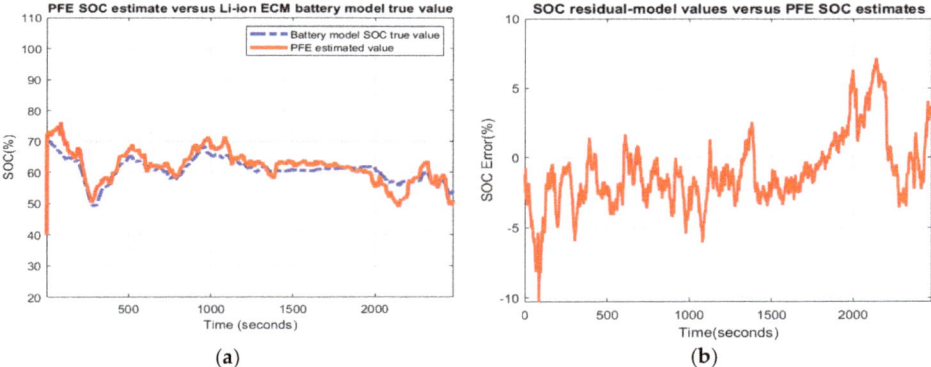

Figure A25. Robustness to changes in SOC initial value—SOCini = 0.4. (**a**) PFE SOC estimate versus 3RC ECM Li-ion battery SOC true value; (**b**) SOC residual.

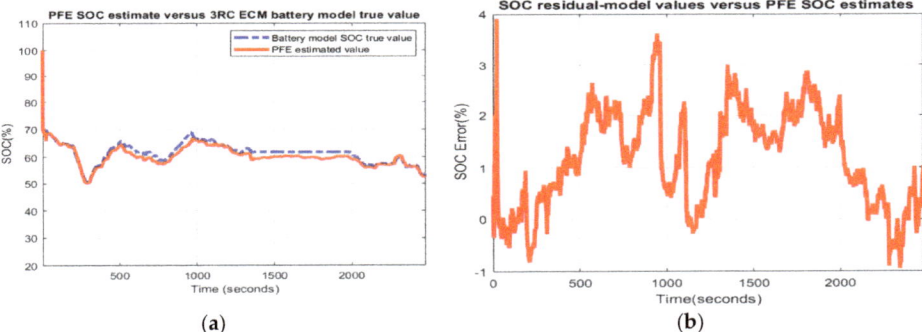

Figure A26. Robustness to simultaneous changes, SOCini = 0.4, σ noise = 0.01; (**a**) PFE SOC value versus 3RC ECM Li-ion battery model true value; (**b**) SOC residual.

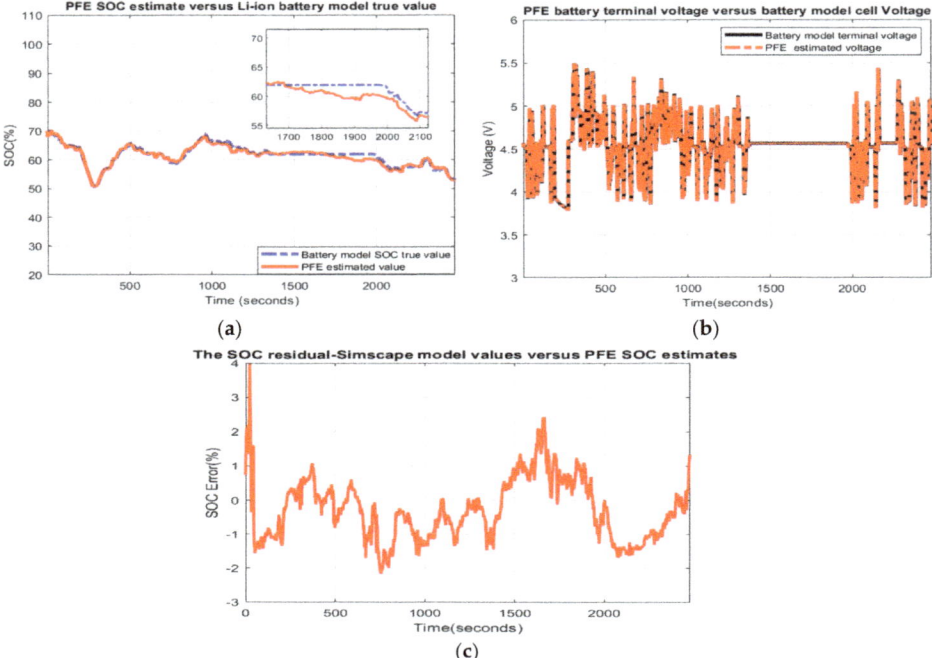

Figure A27. The MATLAB simulation results for Simulink Simscape Li-ion battery. (**a**) The PFE SOC value versus battery model SOC true value; (**b**) The PFE terminal output voltage versus battery model terminal output voltage true value; (**c**) The SOC residual.

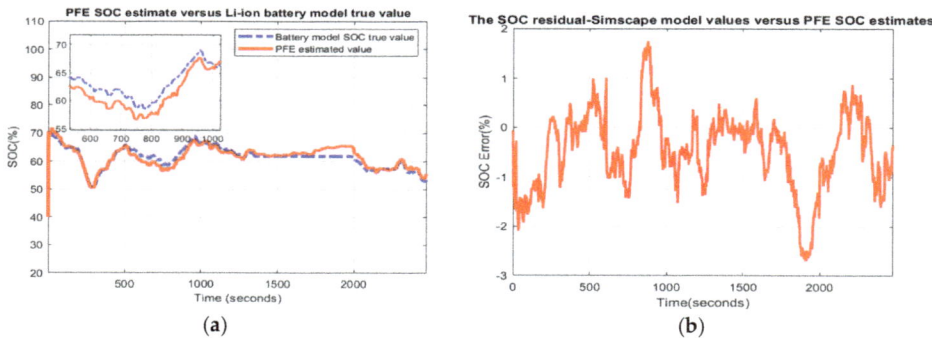

Figure A28. Robustness to changes in SOC initial value—SOCini = 0.4. (**a**) PFE SOC estimate versus battery SOC true value; (**b**) SOC residual.

- Scenario R3: Robustness to simultaneous changes, namely in SOCini (SOCini = 0.4), and measurement noise level, e.g., an increase in noise level 10 times (σ noise = 0.01).

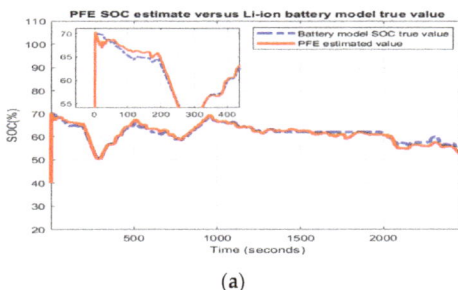
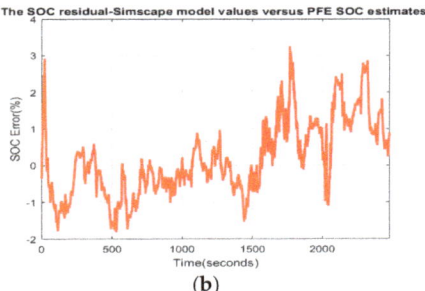

(a)   (b)

Figure A29. Robustness to simultaneous changes, SOCini = 0.4, σ noise = 0.01; (**a**) PFE SOC value versus Simscape battery model true value; (**b**) SOC residual.

Appendix A.2. Tables

Table A1. Statistical errors—SOC estimates versus Li-ion battery Simscape model SOC values—Scenario R1 (SOCini = 0.4).

Performance	Li-Ion Battery 3RC ECM σ = 0.03713			Li-Ion Battery Simulink Simscape Model, σ = 0.036248		
	AEKF	AUKF	PF	AEKF	AUKF	PF
RMSE	0.0101	0.0289	0.0284	0.0185	0.0283	0.0163
MSE	1.0245	0.00008	0.0008	0.00034	0.0008	0.0002
MAE	0.0062	0.00111	0.0226	0.0666	0.0152	0.0119
Standard deviation (σ)	0.0431	0.0433	0.0524	0.0365	0.044	0.0379
MAPE (%)	1.034	2.01	1.4	1.156	2.71	1.89
R^2	0.948	0.636	0.404	0.737	0.598	0.8019

Table A2. Statistical errors—SOC estimates versus Li-ion battery Simscape model SOC values—Scenario R2: (SOCini = 1, Qnom = 4.2 Ah).

Performance	Li-Ion Battery 3RC ECM σ = 0.03713			Li-Ion Battery Simulink Simscape Model σ = 0.036248		
	AEKF	AUKF	PF	AEKF	AUKF	PF
RMSE	0.0117	0.0284	0.047	0.00248	0.0445	0.0146
MSE	0.000137	0.00008	0.00221	6.2e-6	0.00198	0.0002
MAE	0.0099	0.0203	0.03998	0.0101	0.0319	0.112
Standard deviation (σ)	0.0439	0.0728	0.0659	0.0552	0.067	0.0433
MAPE (%)	1.61	3.78	2.05	1.558	4.91	1.77
R^2	0.930	0.6483	0.286	0.3858	0.013	0.840

Table A3. Statistical errors—SOC estimates versus Li-ion battery Simscape model SOC values—Scenario R3: (SOCini = 0.4, σ noise = 0.01).

Performance	Li-Ion Battery 3RC ECM σ = 0.03713			Li-Ion Battery Simulink Simscape Model, σ = 0.036248		
	AEKF	AUKF	PF	AEKF	AUKF	PF
RMSE	0.0812	0.0289	0.016	0.0188	0.0252	0.012
MSE	0.0066	0.0008	0.00221	0.00035	0.00063	0.0001
MAE	0.0728	0.0109	0.3998	0.0029	0.00879	0.0083
Standard deviation (σ)	0.0457	0.0433	0.0659	0.0379	0.047	0.0366
MAPE (%)	10.55	1.9898	2.05	0.52	1.578	1.35
R^2	−2.34	0.637	0.286	0.729	0.681	0.892
Remark	Fail the Test			Pass the Test		

Table A4. Statistical errors—SOC estimates versus Li-ion battery Simscape model SOC values—Scenario R4: (SOCini = 0.2, temperature effects on Rin and Kp).

Performance	Li-Ion Battery 3RC ECM σ = 0.03713			Li-Ion Battery Simulink Simscape Model, σ = 0.036248		
	AEKF	AUKF	PFE	AEKF	AUKF	PFE
RMSE	0.0789	0.042	0.0866	0.0267	0.036	0.0211
MSE	0.0062	0.0018	0.0075	0.000714	0.0012	0.0004
MAE	0.0687	0.0159	0.0668	0.0074	0.0092	0.0167
Standard deviation (σ)	0.0525	0.0464	0.0974	0.04	0.052	0.0337
MAPE (%)	10.5	3.19	3.87	1.44	2.08	1.01
R^2	−2.155	0.233	−4.2	0.456	0.352	0.666

References

1. Farag, M. Lithium-Ion Batteries, Modeling and State of Charge Estimation. Master's Thesis, McMaster University of Hamilton, Hamilton, ON, Canada, 2013.
2. Plett, G.L. Extended Kalman filtering for battery management systems of LiPB-based HEV battery packs: Part 3. State and parameter estimation. *J. Power Sources* **2004**, *134*, 277–292. [CrossRef]
3. Tudoroiu, R.-E.; Zaheeruddin, M.; Radu, S.-M.; Tudoroiu, N. Real-Time Implementation of an Extended Kalman Filter and a PI Observer for State Estimation of Rechargeable Li-Ion Batteries in Hybrid Electric Vehicle Applications—A Case Study. *Batteries* **2018**, *4*, 19. [CrossRef]
4. Tudoroiu, R.-E.; Zaheeruddin, M.; Radu, S.M.; Tudoroiu, N. New Trends in Electrical Vehicle Powertrains-Chapter 4. In *New Trends in Electrical Vehicle Powertrains*, 4th ed.; Martinez, L.R., Prieto, M.D., Eds.; IntechOpen: London, UK, 2019. [CrossRef]
5. Tudoroiu, N.; Zaheeruddin, M.; Tudoroiu, R.-E. Real Time Design and Implementation of State of Charge Estimators for a Rechargeable Li-ion Cobalt Battery with Applicability in HEVs/EVs-A comparative Study. *Energies* **2020**, *13*, 2749. [CrossRef]
6. Zhang, R.; Xia, B.; Li, B.; Cao, L.; Lai, Y.; Zheng, W.; Wang, H.; Wang, W. State of the Art of Li-ion Battery SOC Estimation for Electrical Vehicles. *Energies* **2018**, *11*, 1820. [CrossRef]
7. Simon, J.J.; Uhlmann, J.K. A New Extension of the Kalman Filter to Nonlinear Systems. In Proceedings of the SPIE 3068, Signal Processing, Sensor Fusion, and Target Recognition VI, Orlando, FL, USA, 28 July 1997; p. 3068. Available online: https://people.eecs.berkeley.edu/~{}pabbeel/cs287-fa09/readings/JulierUhlmann-UKF.pdf (accessed on 21 January 2018).

8. Ge, B.; Zhang, H.; Jiang, L.; Li, Z.; Butt, M.M. Adaptive Unscented Kalman Filter for Target Tracking with Unknown Time-Varying Noise Covariance. *Sensors* **2019**, *19*, 1371. [CrossRef] [PubMed]
9. Han, J.; Song, Q.; He, Y. Adaptive Unscented Kalman Filter and its Applications in Nonlinear Control. In *Kalman Filter: Recent Advances and Applications*; Moreno, V.M., Pigazo, A., Eds.; I-Tech: Vienna, Austria, 2009; pp. 1–24.
10. Tudoroiu, N.; Radu, S.M.; Tudoroiu, R.-E. *Improving Nonlinear State Estimation Techniques by Hybrid. Structures*, 1st ed.; LAMBERT Academic Publishing: Saarbrucken, Germany, 2017; p. 56. ISBN 978-3-330-04418-0.
11. Arulampalam, M.S.; Maskell, S.; Gordon, N.; Clapp, T. A tutorial on particle filters for online nonlinear/non-Gaussian Bayesian tracking. *IEEE Trans. Signal Process.* **2002**, *50*, 174–188. [CrossRef]
12. Alvarez, J.M.; Sachenbacher, M.; Ostermeier, D.; Stadlbauer, H.J.; Hummitzsch, U.; Alexeev, A. *Analysis of the State of the Art on BMS*; Everlasting D6.1 Report; Lion Smart GmbH: Munchen, Germany, 2017.
13. Chang, W.-Y. The State of Charge Estimating Methods for Battery: A Review. *ISRN Appl. Math.* **2013**, *2013*, 953792. [CrossRef]
14. Lee, S.J.; Kim, J.H.; Lee, J.M.; Cho, B.H. The state and parameter estimation of an Li-Ion battery using a new OCV-SOC concept. In Proceedings of the 2007 Power Electronics Specialists conference, Orlando, FL, USA, 17–21 June 2007; pp. 2799–2803.
15. He, H.; Liu, Z.; Hua, Y. Adaptive Extended Kalman Filter Based Fault Detection, and Isolation for a Lithium-Ion Battery Pack. *Energy Procedia* **2015**, *75*, 1950–1955. [CrossRef]
16. Zhao, Y.; Xu, J.; Wang, X.; Mei, X. The Adaptive Fading Extended Kalman Filter SOC Estimation Method for Lithium-ion Batteries. *Energy Procedia* **2018**, *145*, 357–362. [CrossRef]
17. Feng, K.; Huang, B.; Li, Q.; Yan, H. Online Estimation of battery SOC for Electric Vehicles Based on An Improved AEKF. *E3S Web Conf.* **2019**, *118*, 02025. [CrossRef]
18. Ma, M.; Qiu, D.; Tao, Q.; Zhu, D. Sate of Charge Estimation of a Lithium Ion Battery Based on Adaptive Kalman Filter Method for Equivalent Circuit Model. *Appl. Sci.* **2019**, *9*, 2765. [CrossRef]
19. Cui, X.; Shen, W.; Zhang, Y.; Cungang, H. A Novel Active Online State of Charge Based Balancing Approach for Lithium-Ion Battery Packs during Fast Charging Process in Electric Vehicles. *Energies* **2017**, *10*, 1766. [CrossRef]
20. Tudoroiu, R.-E.; Zaheeruddin, M.; Tudoroiu, N.; Radu, S.M. SOC Estimation of a Rechargeable Li-Ion Battery in fuel-Cell Hybrid Electric Vehicles-Comparative Study of Accuracy and Robustness performance Based on Statistical Criteria. Part I: Equivalent Models. *Batteries* **2020**, *6*, 40.

© 2020 by the authors. Licensee MDPI, Basel, Switzerland. This article is an open access article distributed under the terms and conditions of the Creative Commons Attribution (CC BY) license (http://creativecommons.org/licenses/by/4.0/).

Article

Experimental Data Comparison of an Electric Minibus Equipped with Different Energy Storage Systems

Fabio Cignini [1,*], Antonino Genovese [1], Fernando Ortenzi [1], Adriano Alessandrini [2], Lorenzo Berzi [3], Luca Pugi [3] and Riccardo Barbieri [3]

1. Italian National Agency for New Technologies and Environment (ENEA), 00123 Rome, Italy; antonino.genovese@enea.it (A.G.); fernando.ortenzi@enea.it (F.O.)
2. Department of Civil and Environmental Engineering (DICEA), University of Florence (UNIFI), 50139 Florence, Italy; adriano.alessandrini@unifi.it
3. Department of Industrial Engineering (DIEF), University of Florence (UNIFI), 50139 Florence, Italy; lorenzo.berzi@unifi.it (L.B.); luca.pugi@unifi.it (L.P.); riccardo.barbieri@unifi.it (R.B.)
* Correspondence: fabio.cignini@enea.it; Tel.: +39-06-3048-3996

Received: 6 April 2020; Accepted: 23 April 2020; Published: 28 April 2020

Abstract: As electric mobility becomes more important every day, scientific research brings us new solutions that increase performance, reduce financial and economic impacts and increase the market share of electric vehicles. Therefore, there is a necessity to compare technical and economic aspects of different technologies for each transport application. This article presents a comparison of three bus prototypes in terms of dynamic performance. The analysis is based on the collection of real data (acceleration, maximum speed and energy consumption) under different settings. Each developed prototype uses the same bus chassis but relies on different energy storage systems. Results show that the dynamic bus performance is independent on the three energy storage technologies, whereas technologies affect the management costs, charging time and available range. An extensive experimental analysis reveals that the bus equipped with a hybrid storage (lithium-ion batteries and supercapacitors) had the most favorable net present value, in comparison with storage composed of only lead–acid or lithium-ion batteries. This result is due to the greater life of lithium-ion batteries and to the capability of supercapacitors, which reduce both batteries depth of discharge and discharge rate.

Keywords: battery; ultracapacitor; supercapacitor; electric mobility; electric bus

1. Introduction

Recent developments in energy storage systems (ESS) and fast charging technologies extend the range of electric vehicles and their increasing market share are reducing prices [1–3].

The European Union set the target of 40% reduction of greenhouse gas (GHG) emissions and of 27% share of renewable energy by 2030 [4], with a potential reduction of 80%–95% of GHG and 55%–75% of gross final energy consumption from renewable sources [5] by 2050.

The transport sector contributes almost a quarter of Europe's GHG emissions and buses are responsible for 8% of transport emissions. In 2019, electric buses all over Europe count 2200 units [6], less than 1% of European bus fleet (about 770,000 units [7]). A study forecasts that electric buses will reach more than 23,000 units in 2025 [8].

An opportunity to shift towards electric transportation is the retrofit [9]. This was fostered in Italy by a recent national policy initiative. The Italian Ministry Economic Development (MISE) issued the order no. 219 of 1 December 2015 [10] to allow this procedure.

A retrofit replaces an Internal Combustion Engine (ICE) with an electric kit (composed by a motor, a battery and some electronics). Today, it is applicable only to M1 and N1 vehicle categories (cars weighing up to 3.5 tons), however there are future possibilities to extend this opportunity to larger vehicles.

This study focuses on small public transport vehicles; those minibuses are maximum 6 m long with a passenger capacity of 30 people. It shows a comparison of the data gathered by three consecutive projects all founded by MISE in the last four years. Project partners were ENEA and four Italian Universities (University of Firenze, Sapienza of Rome, Roma Tre and Pisa).

Each project used the same bus model, a Tecnobus Gulliver ESP500, equipped with different prototypes of energy storage systems (ESS). Figure 1 shows the buses of the three projects:

- Project 1 (P1): Bus equipped with lead–acid batteries, as provided by the original equipment manufacturer (OEM). The project tested an innovative on-demand transport service [11].
- Project 2 (P2): lithium-ion iron phosphate batteries designed for 3C (three times the nominal capacity C) fast charging.
- Project 3 (P3): Hybrid storage with supercapacitors (SC) and absorbent glass mat (AGM) lead batteries, with a flash charging system.

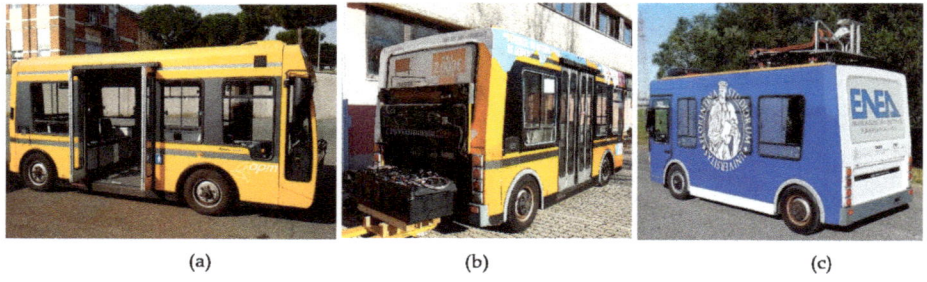

Figure 1. Project prototypes: P1 (**a**), P2 (**b**) and P3 (**c**).

Each project had its own experimental testing campaign, each taking place into the ENEA campus. The campus has about ten kilometers of internal roads, so there are many workers moving during the day. This allowed us simulating a common transport service.

Preliminary results have shown that each bus has same performance and the energy consumption of the vehicle is not influenced by the energy storage. Moreover, P3 achieved better economic results with lower costs. The authors argued the economic benefit is due to less frequent replacements of the battery in a hybrid ESS. Hence, it was evaluated whether it is convenient to replace battery with a more expensive lithium battery. A new project starting from P3 and replacing the AGM battery with a lithium-ion one was simulated and indicated as P4 in the study.

An economical comparison was carried out among these four alternatives: the three experimental projects with real data plus the simulated scenario. It was hypothesized that all prototypes must supply the same transport service with a daily range of 100 km. The comparison of the projects is based on life cycle cost with the net present value (NPV) indicator [12]. The incomes are the same for all of them and costs change from one to another. The project with lower costs has a better NPV.

The present study demonstrates that the best results are achieved in P4, which is characterized by hybrid ESS (as the one of P3) combined with lithium-ion batteries (as the one of P2). P4 combines two technologies with a more efficient usage that give longer life expectation to the electric and storage components.

This study is organized in four sections: the current introduction; Section 2, which presents the details of compared projects; Section 3, which presents the results; and Section 4, the conclusions.

2. Details of the Compared Projects

This section details nominal specifications and analysis of real data for the three projects (from 1 to 3). P4 combines specifications of P2 and P3 and its input data are explained in Section 3.

Project 1 had the lead–acid batteries provided by the OEM. The batteries have no sensors, but the chopper (DC regulator) of the motor provides upon request voltage and current to and from the batteries.

Energy storage system prototypes were manufactured, specifically in P2 and P3. In P2, a battery management system (BMS) measures voltage, current and temperature of each battery-cell individually for safety and advanced management. In P3, a buck-boost DC-DC converter was inserted between the ultracapacitors and the batteries. Therefore, P2 and P3 had very accurate measurement systems installed directly on the energy storage.

These three prototypes have different measurement systems. The first had a 1 Hz sampling rate; the traveled distance was about 100 km. P2 and P3 had an acquisition rate, respectively of 2 Hz and 10 Hz. These two projects had less data and the traveled distance was about 20 km each.

Energy consumption was compared by observing the average consumption per kilometer of many trips, each with different driving cycles, terrain orography, payload and driving styles.

Range and charging times are also different from among projects. In order to be compared and fulfil the same transport service, they required some adaptation:

- P1 had not enough range for a day, considering a typical transport line of 100 km; this required two battery packs and two chargers for each bus. A full charge of battery required eight hours and it covered only about four hours of service, so at midday, driver went to depot, where battery was swapped with a second one (just fully charged).
- P2 had a fast charging feature that allowed for twenty-minute ride followed by seven-minute stop (or forty-minute ride and fourteen-minute stop). The long stop was necessary to charge battery.
- P3 had a flash charging feature that allowed to charge SC in thirty seconds, so it could be charged during transport service stops (while passengers are on board), but it had to happen every 700 m [13].
- P4 has a SC with the flash charging feature, as for P3 and, a LiFePO4 battery (as for P2 but with lesser energy stored).

The choice of LiFePO4 is due to the availability of experimental data [14], where it was estimated the maximum life cycle of a battery with conditions comparable to current bus usage.

A performance and economic evaluation were done. The first one is based on maximum speed, maximum acceleration and time to reach maximum speed with a standing start.

Indeed, the economic evaluation is a cost benefit analysis using the net present value as main indicator of economic value, was performed over a twenty-four-year time frame, to consider a least common multiple of the lifetime expectations of the different technologies (called also cycle life, CL) [15,16].

The periodic replacement of exhausted batteries during the lifetime of the bus has also been considered. An ESS lasts up to a few years depending on the usage. Bus lifetime ranges from 10 to 15 years depending on its size, for example a 12-m long bus has 15 years of depreciation in Italy. The bus used in this analysis is 6-meter long and a lifespan of 12 years was assumed. SC lifetime is longer than a million cycles (according to manufacturer specifications) [17,18].

The bus lifetime was assumed 12 years as reference value, so, the economic evaluation expects at least 24 years, considering at least a replacement for the bus and all ESS components. The SC lifetime is more than twenty years, considering the expected life cycle and their usage in a bus, while, battery life depends on several factors [19]. The main factors are: depth of discharge (DoD), discharge rate (measured in multiples of the nominal capacity C), charge rate, aging, working and environmental temperatures.

For the tests of the all projects, the working temperature of the battery was maintained within the limits prescribed by manufacturer. While the environmental temperatures of these projects were the same, all tests made in the ENEA campus were conducted with mild weather.

It can be assumed that for a bus application, number of cycles life of ESS components is lower than the calendar life (it is the elapsed time before a battery becomes unusable, whether it is in active or inactive use). The manufacturer of the AGM battery, used in P1 and P3, declares 20 months of life, while lithium lasts up to five years [20,21].

Meanwhile, DoD has the highest impact to a battery; for the P1 equipped with a lead–acid battery, the DoD was about 80% and it had 500 cycles to failure [22], as shown in Figure 2. Hence, if the transport service application requires a full charge every day, the battery must be replaced every 500 days.

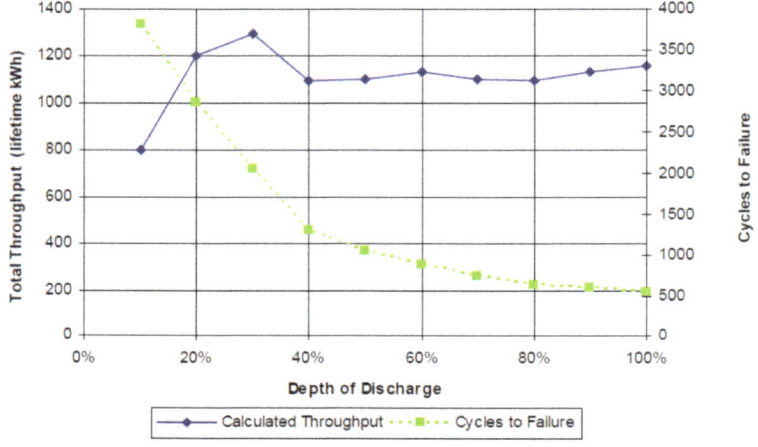

Figure 2. Lead–acid battery behavior [22].

Figure 3 shows DoD effects applied to three lithium batteries with different chemistry [23]. A LiFePO4 battery (as those of P2) has about a thousand CL if used up to 80% of DoD or, if used only up to 40%, it will last three times longer.

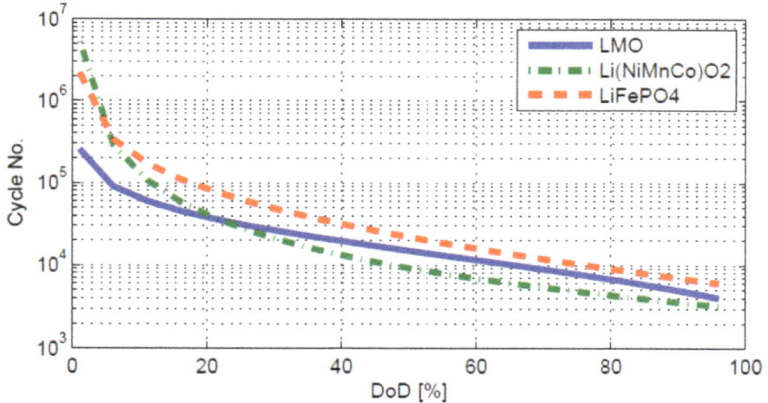

Figure 3. Effects of DoD to a few types of lithium battery [23].

Such effects become very important from an economic point of view especially in the case of higher costs of lithium compared with lead–acid.

C rate, during charging and discharging, reduces the battery lifetime even more. As described in Figure 4, the capacity of a lead–acid battery drops when discharge rate raises from 0.5C to 10C, then battery capacity decreases from 100% of initial value (battery fully charged) to 70% [24]. Lithium-ion batteries suffer from the same issues but have different effects [14,25], accordingly to the discharge rate capability and battery life cycle given by battery manufacturers.

Experimental results of three cited project are described one by one as follows.

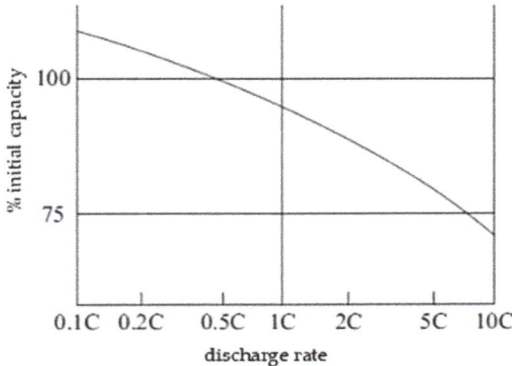

Figure 4. Rate capacity effect [24] with a lead–acid battery.

2.1. Project 1 (P1): Lead–acid Batteries

The goal of the project was to develop an on-demand transport system between ENEA facilities. The bus was equipped with a lead–acid battery of 43 kWh at 72 V, with a capacity of 600 Ah. It was composed by 36 batteries of 100 Ah–12 V each. The configuration consisted of 2 strings in series, each one composed of eighteen batteries, meaning six groups in parallel of three batteries in series.

It had long running acquisition including different missions. Each one of them included a running distance of at least 500 m. It started and stopped at zero speed (minimum measuring time of 10 s).

This project ran for almost three months and data for about 100 kms were collected. In order to evaluate such amount of data, focusing only on the average consumption and with a wide variety of driving conditions (slopes, payload, etc.), data were divided into more than 110 stretches. Figure 5 shows the histogram of the occurrences for the average consumption.

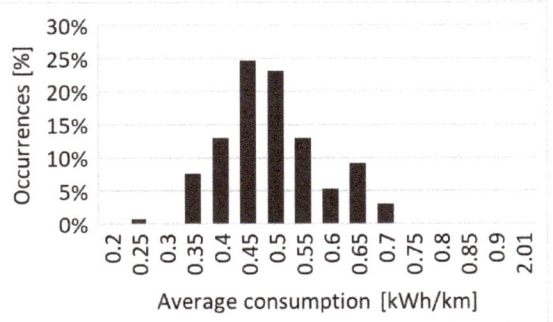

Figure 5. Occurrences of average consumption for different stretches.

This bus consumes from 0.35 to 0.70 kWh/km, with a modal value of 0.45 kWh/km.

Figure 6 shows the current and voltage trend during a stretch. The ESS provided more than 320 A and each battery up to 55 A of maximum current. The chopper of P1 did not show negative values through the interface used, but it also computed energy consumption with both negative and positive currents.

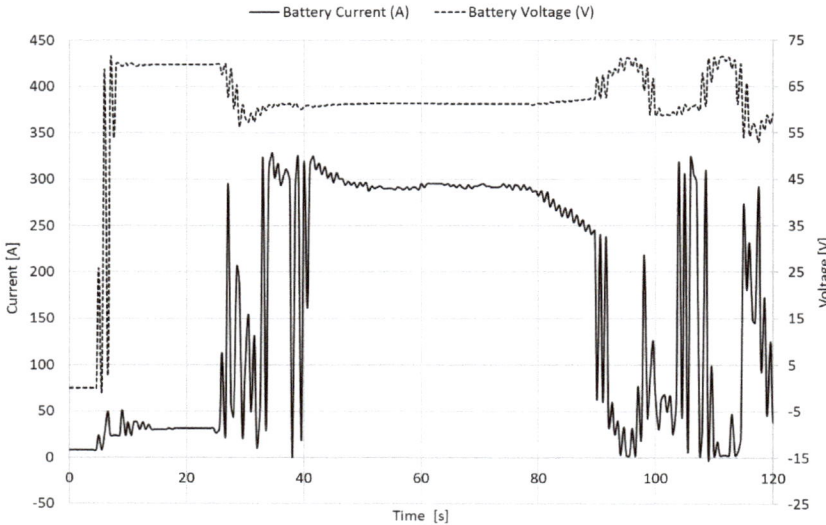

Figure 6. Voltage and current trend for P1 during a bus ride.

Figure 6 highlights the voltage of battery drops due to high internal resistances.

The internal resistance further reduces the life expectation of lead–acid batteries. A new battery features this behavior only with low state of charge (SOC), but it gets worse with age and number of cycles.

Table 1 shows the results of a few rides during transport service of P1.

Table 1. Road testing results of P1.

Parameters	Unit	1	2	3	4	5	Average	Total
Duration	s	241.0	491.0	122.0	438.5	576.0	388.7	68,411
Distance	km	1.08	1.68	0.68	2.49	2.97	1.0	176
Total consumption	Wh	432.3	846.0	286.8	942.9	1,218	491.3	86,463
Commercial speed	km/h	16.2	12.3	19.9	20.4	18.5	11.6	n.a.
Average consumption	Wh/km	398.5	503.3	424.0	378.9	410.6	489.4	n.a.

2.2. Project 2 (P2): Lithium-Ion Batteries

This project developed a prototype of fast charging battery pack for a small minibus [26,27].

The prototype battery pack was composed of 17 kWh of lithium batteries. It is composed of 96 cells of 3.7 V and 60 Ah each. The configuration was four strings of twenty-four cells in series each and the whole battery reached 76 V–240 Ah. It was capable of 3C charging rate.

The chemical composition was LiFePO4; it could be charged with 1.4 kWh in 110 s as shown in Figure 7, where current and energy during fast charging are plotted. Current values of Figure 7 are negative due to sign convention of the measurement system. Hence, the energy decrease means a charge.

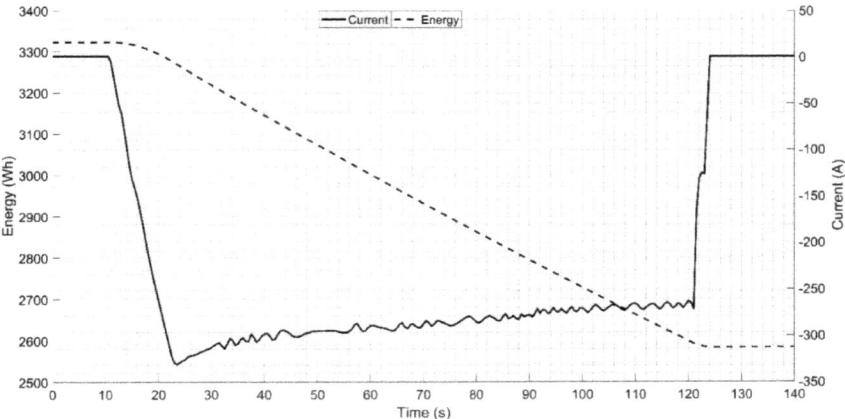

Figure 7. Battery energy and current trends during charging.

Figure 8 shows voltage and current trend during a bus ride.

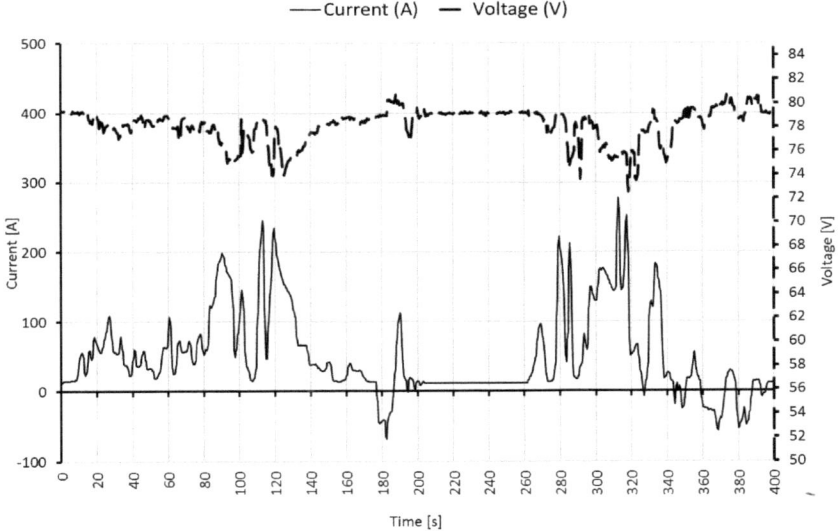

Figure 8. Voltage and current trends of P2.

There are two important differences between P1 and P2:

- Maximum discharge current of whole battery pack: the lead–acid one had a greater current (330 A against 280 A). Thanks to parallel and series connections the maximum current of whole ESS was divided among batteries, so each lead–acid ESS achieved up to 50 A of maximum current, while lithium-ion achieved up to 70 A.
- Voltage range: The lead–acid battery had about 20 V (from 53 V to 73 V), whereas the lithium-ion battery had 8 V (from 73 V to 81 V)

The lowest value of minimum voltage is probably due to high-power-request battery with degraded state of health or even low levels of SOC. These low-voltage situations cause malfunctioning in auxiliary devices (i.e., DC–DC converters, steering pump, brake pump, relays, etc.) and increase currents.

Such situation starts a chain reaction that alters the battery composition. A given power request with lower voltage means higher current (in comparison with another one at higher voltage). The higher current, in turn, means higher losses in heating and further lower voltage (due to rise of internal resistances) and again much higher current.

Table 2 shows the results of a few rides for P2; that needs 422 Wh of energy per kilometer.

Table 2. Road testing results of P2.

Parameters	Unit	1	2	3	4	5	Average *	Total *
Duration	s	440	440	405	640	529	491	2454
Distance	m	1104	1599	1147	1180	1340	1274	6370
Total consumption	Wh	574	510	435	550	557	535	2673
Commercial speed	km/h	9.0	13.1	10.2	6.6	9.1	9.3	n.a.
Average consumption	Wh/km	511	313	364	441	396	422	n.a.

*: based on all data measured.

2.3. Project 3 (P3): Hybrid Storage SC and AGM Batteries

Figure 9 shows the prototype while it charged at bus stop. The project design was published [28–30].

Figure 9. Charging phase of the prototype of P3 with flash charging technology.

The bus was equipped with a hybrid storage system composed of AGM batteries and supercapacitors. The goal of this project was to develop a flash charging technology for public transport that can charge small quantities of energy very quickly at every stop.

The SC provides through the DC–DC converters some energy directly to the chopper, reducing the energy provided by the battery [31,32].

Figure 10 shows voltage and current trends during charging. This phase lasts 45 s and charges up to 302 Wh, current reaches 350 A. Supercapacitor voltage ranges from 200 V to 375 V. These features can be further improved by optimizing the charging phase; after some tests, an optimistic hypothesis is 20 s (to be validated).

This project requires a charging station at least every 600 m. Larger distances between two charging station can deplete the energy stored on board [33].

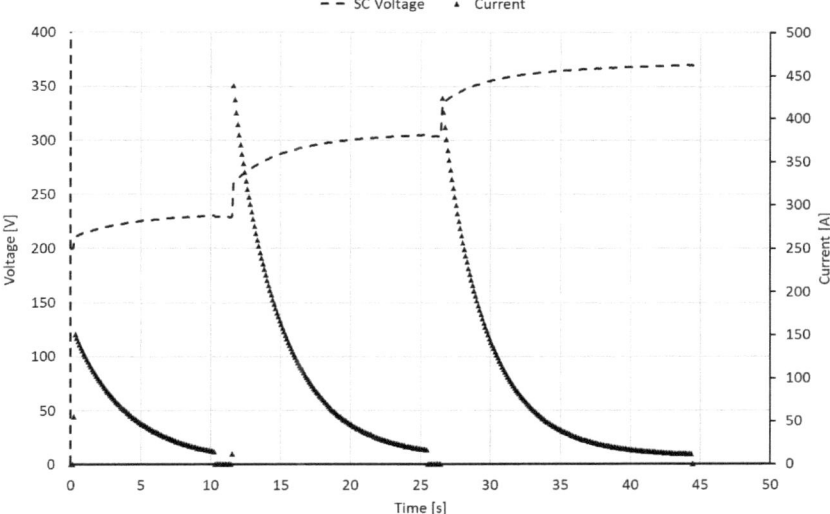

Figure 10. Supercapacitor voltage and current trends during the ultra-fast charging phase.

Table 3 shows main results of experimental campaign, it also shows more parameters than previous projects due to a larger number of installed sensors (i.e., DC–DC converters and SC).

Every hour the control strategy of DC–DC drains 35 Ah from the battery, so this prototype guarantees an autonomy of 3 h, or about 35 km, if the battery capacity is 120 Ah as in the P3 prototype. After that, it requires a slow full charge of AGM battery.

Table 3. Main test results of hybrid storage prototype (P3).

Parameter	Unit	1	2	3	4	5	Average	Total
Duration	sec	272.7	263.1	307.5	252.5	250.0	279.2	3629.3
Distance traveled	km	1.15	0.96	1.00	0.58	0.86	0.9	11.5
Total energy consumption	Wh	515.5	487.6	421.9	325.1	416.4	457.7	5950.5
Energy consumption due to auxiliary services	Wh	83.4	79.5	92.9	77.7	76.4	84.8	1102.9
Traction motor energy consumption	Wh	432.1	408.1	329.0	247.4	340.1	372.9	4847.6
Average energy consumption	Wh/km	447.7	509.8	422.1	562.7	483.5	528.0	n.a.
Commercial speed	km/h	15.2	13.1	11.7	8.2	12.4	11.7	n.a.
Energy provided by battery	Wh	266.5	241.2	175.6	131.3	152.5	208.4	2709.1
Energy provided by supercapacitors	Wh	249.0	246.3	246.2	193.8	263.9	249.3	3241.4
Supercap rate usage	%	58%	60%	75%	78%	78%	68%	n.a.
Battery Ah	Ah	3.5	3.2	2.3	1.7	2.0	2.7	35.6

This project tested only a few strategies to manage the hybrid storage; a next step will be to reduce the current drained from the battery in order to increase its life.

Further developments come from using different strategies in order to keep the energy provided by battery close to zero, for example a depleting strategy allows to fulfil daily mileage required.

Figure 11 shows that the maximum current drained from the hybrid storage is about 320 A, 100 A of which are provided by the battery. The maximum current provided by the battery is the main difference between P3 and previous P1 and P2. Moreover, it brings a great benefit to lead–acid battery that has less voltage fluctuation (between 69 V and 78 V) than P1.

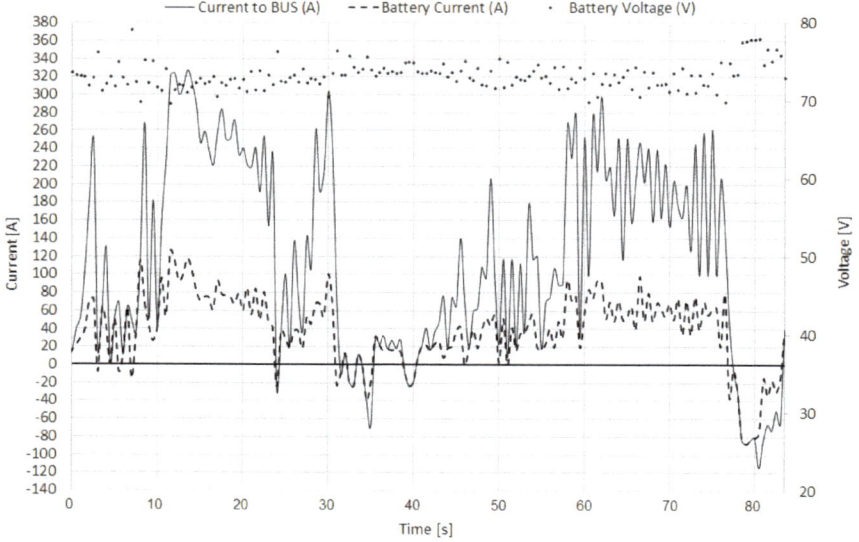

Figure 11. Battery voltage and current trends.

3. Results Comparison

This chapter compares the results of the three described energy storage systems installed on the same bus, with nominal specification described in Table 4. The comparison refers to those systems that offer the same transport service of 100 km per day (300 day per year). The bus powertrain remains the same, the traction motor requires 25 kW during peak request and 20 kW continuously.

Table 4. Bus nominal specifications.

Nominal Parameter	Unit	Value
Purchase cost (bus without battery)	€	200,000
Lifespan	year	12
Average consumption (OEM data)	kWh/km	0.5
Daily mileage required (min)	km	100
Curb weight (without battery)	kg	2370
Motor power (DC-brushed)	kW	21 (25 peak)
Motor torque	Nm	235 (at 950 RPM)

Each prototype had a different weight, due to different technology installed, from batteries (SC if present), BMS, mechanical supplementary frame, pantograph, to additional electronics, etc. Weights are detailed in Table 5. Such differences of weight could affect performances. Hence, it was conducted tests of maximum speed, maximum acceleration and time from zero to maximum speed. The results indicate that there were no relevant differences, all buses reach 33 km/h in 60 s with a maximum acceleration of 0.6 m/s^2.

Table 5. ESS weight in the projects.

Component	Description	Unit	Project 1	Project 2	Project 3	Project 4
			Lead–acid	LiFePO4	SC + PbAGM	SC + Lithium
ESS weight	Battery, electronics and other mechanical additions total weight	kg	1500	800	1100	700

The comparison shown in Figure 12 allows to assume the same average consumption, independently from technology. Due to minimum differences measured during tests, it could be different due to phenomena such as orography, payload, driving behavior, etc. Hence, the adoption of the same energy consumption value is a conservative choice. Figure 12 shows the average consumption versus the average speed trend, during different missions for the three projects P1, P2 and P3. The cloud is denser for P1, due to the large amount of data.

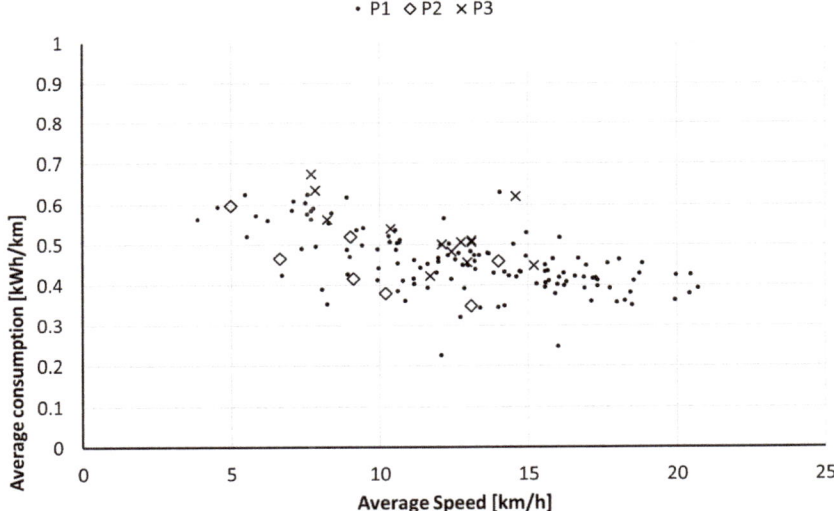

Figure 12. Trend of energy consumption related to average speed.

A slight difference in average consumption, e.g., of about 0.1 kWh/km (20% of average consumption), corresponds to consumption of 10 kWh per day and 3000 kWh per year.

The average energy cost depends on market factors and power requirements; in Italy, this value is between 0.1 and 0.3 Euro per kWh, leading to a cost of 300 to 900 Euro per year.

Hence, the economic value of energy is negligible in comparison with the battery itself.

Table 6 shows a comparison of parameters (partly proposed by [34]), costs and results.

Table 6. Comparison of cost parameters.

Component	Parameter	Unit	P1 lead–acid	P2 LiFePO4	P3 SC + PbAGM	P4 SC + Li
Battery	Lifecycle	#	800	3000	3000	9000
	Energy	kWh	43.2	17.3	8.6	4.3
	Unitary cost	€/kWh	150	450	150	450
	Range with a single charge	km	67.2	30.7	11.6	5.8
	Total cost	€	12,960	7776	1296	1944
	Number of daily full charge	#	1	3	5	5
	Life	Year	2.7	3.3	2.0	5.0
Supercap	Lifecycle	#	Na	Na	1,000,000	1,000,000
	Energy	kWh	Na	Na	0.4	0.4
	Unitary costs	€/kWh	Na	Na	37,000	37,000
	Range with a single charge	km	Na	Na	0.7	0.7
	Number of daily full charge	#	Na	Na	141	141
	Life	Year	Na	Na	23.6	23.6
	Total cost	€	Na	Na	15,170	15,170
Charger	Cost due to one bus	€	3000	3333	2500	2500
	Charger life	Year	12	12	12	12

Table 6 is divided into main components: battery, supercap and charger; each component is detailed in the following.

Battery parameters are:

- Lifecycle of P1 and P2 comes from project results.
- Lifecycle of P3 is an estimation based on theory [22] and a maximum DoD of 20%.
- Lifecycle of P4 is evaluated with a mixed approach from theory and results obtained in P2. The LiFePO4 battery lasts 3000 cycles at 80% of DoD, P4 uses only the 40% of DoD so according to [23] its battery lasts three times that of the P2.
- Energy is the total energy stored in the ESS. P1 and P2 need the battery to provide power for traction motor and energy for the daily service, whereas P3 and P4 require less battery capacity. In fact, the presence of SC and fast charge feature (P3 and P4) ensure the energy needs, so, battery is dimensioned with at least 20 kW of power (to supply motor request in case of SC failure).
- Unitary costs refer to available products on the market.
- Range with a single charge is the maximum range without intermediate charging.
- Number of daily full charge represents the number of charges needed daily for the required transport service; each charge refills the battery up to daily initial SoC.
- Life is the expected time in years before replacement due to battery usage, it is the lifecycle parameter divided by annual cycles (number of daily full charge per daily of transport service, 300 days per year). But it must not exceed the battery calendar life (as in the P4, the LiFePO4 has 5 years).

Supercap parameters are:

- Lifecycle is provided by manufacturer under nominal working conditions.
- Energy stored is total energy of supercapacitor from maximum voltage to half voltage, as prescribed by manufacturer in order to preserve their life.
- Unitary costs are equal to the costs paid for the prototype of P3; recent updates suggest a reduction down to 32,000 $/kWh [34,35] or even, calculated in Farad, from 1 cent €/F down to 0.1 cent€/F [36]. P3 has three modules in series of SC, each composed by forty-eight cells of 3000 F.
- Range with a single charge is the maximum range, it considers the effective energy (up to half nominal voltage).

Charger parameters are:

- Cost due to one bus: some chargers can be used for many buses. The cost is the same paid for prototypes, but it can be reduced with a large-scale production. In other words, the P3 was equipped with flash charge and it used a charger only for forty seconds, then charger required five minutes more to restore its energy before charging another bus. A single charger costs up to twenty thousand Euros.

The SC adoption in P3 and P4 achieves two great benefits: it reduces the maximum current and the overall energy drained by the battery, during a daily transport service. Such benefits enlarge the number of battery cycles, while reduce the amount of energy stored on board (it decreases battery dimension, weight and cost).

Figure 13 shows trends of costs for each project, included the P4 with supercapacitor and lithium batteries (represented by black dots). Figure 13 highlights when hardware replacements will occur (as bus chassis, batteries, etc.) in a twenty-four-year timeframe.

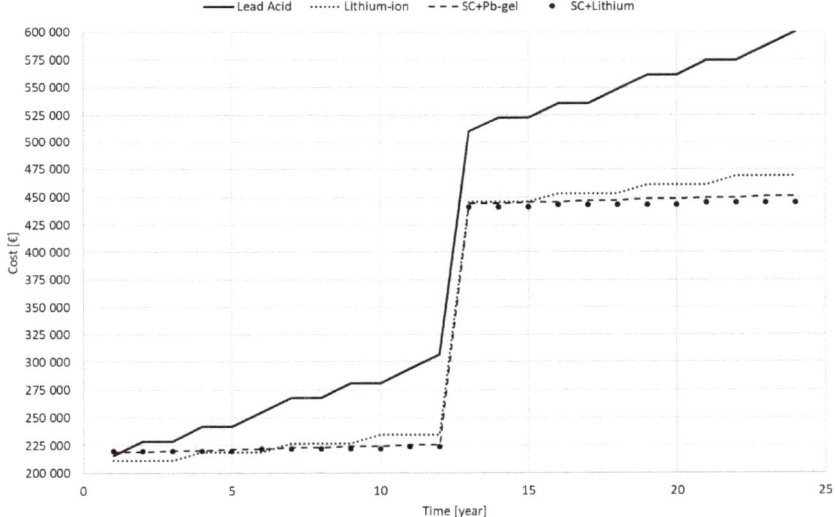

Figure 13. Economic comparison.

All projects have the same incomes, due to the same transport service; only the costs change from one to another. Hence, a lower cost means a better NPV.

Worth of incomes depends on factors as transport policy, economic situation, social aspects, political choices, etc. Thus, calculating the incomes is not useful since they will equally affect all systems.

Table 7 shows results of the sum between the actualized costs for the four alternatives. This parameter is used in the net present value (NPV) evaluation [12], the cost of capital is set to 3%. P4 has lower actualized costs, closer to P3. These two projects have a higher purchase cost due to SC, but they have larger lifetime. However, starting from the seventh year, the sum of costs (as shown in Figure 13) is lower than in P1 and P2.

Table 7. Actualized costs.

Parameter	Unit	Project 1 lead–acid	Project 2 LiFePO4	Project 3 SC + PbAGM	Project 4 SC + Lithium
Actualized Costs	€	487,156	392,504	381,810	377,731

4. Conclusions

This study compares different energy storage systems for electric minibuses. They were prototyped and tested. A cost-benefit analysis was carried out to compare the different solutions from two points of view: economic and performance.

The compared solutions are based on experimental data of four projects (from to P1 to P4). The performance analysis shows that each prototype reaches maximum speed of 33 km/h in 60 s with same maximum acceleration of 0.6 m/s^2.

By economic point of view, the best choice is SC + lithium, which has the lowest actualized costs (so the best NPV), but they are close to SC + Pb. These results can change with fluctuation of product prices. Moreover, expected lifecycle of batteries must be demonstrated under several different conditions that were at this stage hypothesized. Indeed, P2 and P3 have both a large cost reduction compared to P1, so, P4 joins their technical advantages and have at same time a favorable actualized cost.

Technology choices in transport must consider several technical factors, i.e., in case of failure of supercapacitors, the battery must guarantee enough range to reach next stop or even to return to the depot. Other important factors are the capabilities of fast charging and the high power.

Based to the aforementioned considerations indicate P4 as the best option. lithium-ion battery coupled with SC guarantees required energy, sufficient power and highest charging rates.

Further developments may come from new testing campaigns to demonstrate the lifecycle increasing of a LiFePO4 battery combined with SC (as hypothesized in P4)—or even with a new battery chemistry. Experimental counting of the number of cycles will allow a realistic evaluation of battery life and could enrich the current economic evaluation.

Author Contributions: All the authors equally contributed to state-of-the-art survey, prototype project and experiments design. A.G. acquired funding and supervised, A.A. and F.O. managed resources and administrated projects. The data curation was conducted by L.B., R.B., F.C. and L.P. Moreover, L.B., F.O. and L.P. developed software required to manage electronic devices. F.C. together with L.P. and R.B. validated the results. All authors have read and agreed to the published version of the manuscript.

Funding: This research was funded by Italian Ministry for Economic Development (MISE), program agreement MISE-ENEA 2015–2017.

Conflicts of Interest: The authors declare no conflict of interest. The funders had no role in the design of the study; in the collection, analyses or interpretation of data; in the writing of the manuscript or in the decision to publish the results.

References

1. Aneke, M.; Wang, M. Energy storage technologies and real life applications—A state of the art review. *Appl. Energy* **2016**, *179*, 350–377. [CrossRef]
2. Peng, H. *Charging Infrastructures for Electric Buses*; Politecnico di Milano: Milan, Italy, 2019.
3. Korkmaz, E. Technically and Economically Viable Future Electricity and Fuel Storage Technologies. Available online: https://aaltodoc.aalto.fi/handle/123456789/40806 (accessed on 20 January 2020).
4. European Commission. A Policy Framework for Climate and Energy in the Period from 2020 to 2030. Available online: https://eur-lex.europa.eu/legal-content/EN/TXT/?uri=COM%3A2014%3A15%3AFIN (accessed on 20 January 2020).
5. European Commission. A Roadmap for Moving to a Competitive Low Carbon Economy in 2050. Available online: http://www.europarl.europa.eu/meetdocs/2009_2014/documents/com/com_com(2011)0112_/com_com(2011)0112_en.pdf (accessed on 20 January 2020).
6. Green Car Congress. Number of Electric Buses in Europe has Increased from Around 200 to 2200 in 5 years. BusWorld. Available online: https://www.greencarcongress.com/2019/10/20191020-busworld.html (accessed on 20 January 2020).
7. European Automobile Manufacturers' Association (ACEA). ACEA Report: Vehicles in Use. 2019. Available online: https://www.acea.be/uploads/publications/ACEA_Report_Vehicles_in_use-Europe_2019.pdf#page=7 (accessed on 20 January 2020).

8. Quéromès, A.; Vogelaar, M.; Huisman, R. Electric Bus Fleets in Europe. Available online: https://www.accuracy.com/wp-content/uploads/2020/01/Perspectives-Print-Electric-bus-v1_anglais.pdf (accessed on 23 January 2020).
9. Alessandrini, A.; Cignini, F.; Ortenzi, F.; Pede, G.; Stam, D. Advantages of retrofitting old electric buses and minibuses. *Energy Procedia* **2017**, *126*, 995–1002. [CrossRef]
10. Italian Minister for Economic Development (MISE). DECRETO 1 Dicembre 2015, n. 219. 2016. Available online: https://www.gazzettaufficiale.it/eli/id/2016/01/11/15G00232/sg (accessed on 23 January 2020).
11. Parragh, S.N.; Doerner, K.F.; Hartl, R.F. Demand Responsive Transportation. In *Wiley Encyclopedia of Operations Research and Management Science*; Wiley Online Library: Hoboken, NJ, USA, 15 February 2011; Online ISBN 9780470400531.
12. Ricci, S. *Tecnica ed Economia dei Trasporti*; Hoepli: Milano MI, Italy, 1 March 2011.
13. Ortenzi, F.; Pasquali, M.; Prosini, P.P.; Lidozzi, A.; di Benedetto, M. Design and Validation of Ultra-Fast Charging Infrastructures Based on Supercapacitors for Urban Public Transportation Applications. *Energies* **2019**, *12*, 2348. [CrossRef]
14. Pede, G.; Vellucci, F. Fast-Charge Life Cycle Test on a Lithium-Ion Battery Module. *World Electr. Veh. J.* **2018**, *9*, 13.
15. Cantarella, G.E. *Sistemi di Trasporto: Tecnica ed Economia*; UTET: Torino, Italy, 2007.
16. Cartenì, A. *Processi decisionali and Pianificazione dei Trasporti*; Lulu Press, Inc.: Morrisville, NC, USA, 16 September 2016.
17. Sheth, A.; Sarkar, D. Life cycle cost analysis for electric vs diesel bus transit in an indian scenario. *Int. J. Technol.* **2019**, *10*, 105–115. [CrossRef]
18. Potkány, M.; Hlatká, M.; Debnár, M.; Hanz, J. Comparison of the Lifecycle Cost Structure of Electric. *Naše More* **2018**, *65*, 270–275. [CrossRef]
19. Wikner, E. Lithium ion Battery Aging: Battery Lifetime Testing and Physics-based Modeling for Electric Vehicle Applications. Ph.D. Thesis, Chalmers University of Technology, Göteborg, Sweden, 2017.
20. Miranto, A. BU-1003a: Battery Aging in an Electric Vehicle (EV). Battery University. Available online: https://batteryuniversity.com/learn/article/bu_1003a_battery_aging_in_an_electric_vehicle_ev (accessed on 13 February 2020).
21. Ortenzi, F.; Pede, G.; Prosini, P.P.; Andrenacci, N. Ageing effects on batteries of high discharge current rate. In *EVS30 Symposium*; Landesmesse Stuttgart GmbH: Stuttgart, Germany, 2017.
22. Bindner, H.; Lundsager, P.; Cronin, T.; Manwell, J.F.; Abdulwahid, U.; Baring-Gould, I. *Lifetime Modelling of Lead Acid Batteries*; Forskningcenter Risoe: Roskilde, Denmark, 2005.
23. Xu, B.; Oudalov, A.; Andersson, G.; Ulbig, A. Modeling of Lithium-Ion Battery Degradation for Cell Life Assessment. *IEEE Trans. Smart Grid* **2016**, *9*, 1131–1140. [CrossRef]
24. Jongerden, M.; Haverkort, B. *Battery Modeling*; (CTIT Technical Report Series; No. TR-CTIT-08-01); Design and Analysis of Communication Systems (DACS): London, UK, 2008.
25. Yang, X.-G.; Zhang, G.; Ge, S.; Wang, C.-Y. Fast charging of lithium-ion batteries at all temperatures. *Proc. Natl. Acad. Sci. USA* **2018**, *115*, 7266–7271. [CrossRef] [PubMed]
26. Baronti, F.; di Rienzo, R.; Moras, R.; Roncella, R.; Saletti, R.; Pede, G.; Vellucci, F. Implementation of the fast charging concept for electric local public transport: The case-study of a minibus. In Proceedings of the IEEE 13th International Conference on Industrial Informatics (INDIN), Cambridge, UK, 22–24 July 2015.
27. di Rienzo, R.; Baronti, F.; Vellucci, F.; Cignini, F.; Ortenzi, F.; Pede, G.; Roncella, R.; Saletti, R. Experimental analysis of an electric minibus with small battery and fast charge policy. In Proceedings of the 2016 International Conference on Electrical Systems for Aircraft, Railway, Ship Propulsion and Road Vehicles & International Transportation Electrification Conference (ESARS-ITEC), Toulouse, France, 2–4 November 2016.
28. Cignini, F.; Genovese, A.; Ortenzi, F.; Alessandrini, A.; Barbieri, R.; Berzi, L.; Locorotondo, E.; Pierini, M.; Pugi, L. Design of a Hybrid Storage for Road Public Transportation Systems. In *IFToMM ITALY*; Springer: Cham, Switzerland, 2018.
29. Ortenzi, F.; Pasquali, M.; Pede, G.; Lidozzi, A.; Benedetto, M. Ultra-fast charging infrastructure for vehicle on-board ultracapacitors in urban public transportation applications. In *EVS31, EVTeC and APE*; Society of Automotive Engineers of Japan, Inc.: Kobe, Japan, 2018.

30. Cignini, F.; Genovese, A.; Ortenzi, F.; Alessandrini, A.; Barbieri, R.; Berzi, L.; Locorotondo, E.; Pierini, M.; Pugi, L.; Baldanzini, N.; et al. Structural and energy storage retrofit of an electric bus for high-power flash recharge. *Procedia Struct. Integr.* **2019**, *24*, 408–422.
31. Alessandrini, A.; Cignini, F.; Barbieri, R.; Berzi, L.; Pugi, L.; Pierini, M. *Progettazione and Test di un Sistema Ibrido SC-accumulo per la Ricarica Rapida di un bus Alle Fermate*; ENEA: Rome, Italy, 2018.
32. Alessandrini, A.; Cignini, F.; Barbieri, R.; Berzi, L.; Pugi, L.; Genovese, A.; Pierini, M.; Locorotondo, E. A Flash Charge System for Urban Transport. In Proceedings of the International Conference on Environment and Electrical Engineering and 2019 IEEE Industrial and Commercial Power Systems Europe (EEEIC/I&CPS Europe), Genova, Italy, 11–14 June 2019.
33. Ortenzi, F.; Pede, G.; Orchi, S. Technical and economical evalutation of hybrid flash-charging stations for electric public transport. In Proceedings of the 2017 IEEE International Conference on Industrial Technology (ICIT), Toronto, ON, Canada, 22–25 March 2017.
34. Mongird, K.; Fotedar, V.; Viswanathan, V.; Koritarov, V.; Balducci, P.; Hadjerioua, B.; Alam, J. *Energy Storage Technology and Cost Characterization Report*; HydroWIRES; Pacific Northwest National Lab.(PNNL): Richland, WA, USA, 2019.
35. Conte, M.; Ortenzi, F.; Genovese, A.; Vellucci, F. Hybrid battery-supercapacitor storage for an electric forklift: A life-cycle cost assessment. *J. Appl. Electrochem.* **2014**, *44*, 1–10. [CrossRef]
36. Skeleton Technologies GmbH. A new approach for ultracapacitor-battery hybrid energy storage solutions. In *International Workshop on Supercapacitors and Energy Storage*; Skeleton Technologies GmbH: Bologna, Italy, 2019.

 © 2020 by the authors. Licensee MDPI, Basel, Switzerland. This article is an open access article distributed under the terms and conditions of the Creative Commons Attribution (CC BY) license (http://creativecommons.org/licenses/by/4.0/).

Article

Measuring Test Bench with Adjustable Thermal Connection of Cells to Their Neighbors and a New Model Approach for Parallel-Connected Cells

Alexander Fill [1,*], Tobias Mader [1], Tobias Schmidt [1], Raphael Llorente [1] and Kai Peter Birke [2]

1. Research and Development, System Design, Daimler AG, Neue Strasse 95, 73230 Kirchheim unter Teck, Germany; tobias.mader@daimler.com (T.M.); tobias.st.schmidt@daimler.com (T.S.); raphael.llorente_cerdan@daimler.com (R.L.)
2. Electrical Energy Storage Systems, Institute for Photovoltaics, University of Stuttgart, Pfaffenwaldring 47, 70569 Stuttgart, Germany; peter.birke@ipv.uni-stuttgart.de
*   Correspondence: alexander.fill@daimler.com

Received: 28 October 2019; Accepted: 20 December 2019; Published: 26 December 2019

Abstract: This article presents a test bench with variable temperature control of the individual cells connected in parallel. This allows to reconstruct arising temperature gradients in a battery module and to investigate their effects on the current distribution. The influence of additional contact resistances induced by the test bench is determined and minimized. The contact resistances are reduced from $R_{Tab+} = 81.18$ µΩ to $R_{Tab+} = 55.15$ µΩ at the positive respectively from $R_{Tab-} = 35.59$ µΩ to $R_{Tab-} = 28.2$ µΩ at the negative tab by mechanical and chemical treating. An increase of the contact resistance at the positive tab is prevented by air seal of the contact. The resistance of the load cable must not be arbitrarily small, as the cable is used as a shunt for current measurement. In order to investigate their impacts, measurements with two parallel-connected cells and different load cables with a resistance of $R_{Cab+} = 0.3$ mΩ, $R_{Cab+} = 1.6$ mΩ and $R_{Cab+} = 4.35$ mΩ are conducted. A shift to lower current differences with decreasing cable resistance but qualitatively the same dynamic of the current distribution is found. An extended dual polarization model is introduced, considering the current distribution within the cells as well as the additional resistances induced by the test bench. The model shows a high correspondence to measurements with two parallel-connected cells, with a Root Mean Square Deviation (RMSD) of $\xi_{RMSD} = 0.083$ A.

Keywords: lithium-ion battery; parallel-connected cells; measuring test bench; current distribution; tab contact resistance

1. Introduction

Large-scale battery applications like electric vehicles (EV) have to meet high power and energy demands, which is mostly realized by the parallel-connection of lithium-ion cells, e.g., Tesla Model S (74p96s (The abbreviation xpys corresponds to a cell configuration with x cells in parallel and y cells in serial connection)), Tesla Model 3 (46p96s), VW eGolf (3p88s), Nissan Leaf (2p96s), BAIC EU260 (3p90s), Renault Zoe (2p96s) and Audi etron (4p108s) [1,2]. Caused by production-induced distributions of cell resistances and capacities [3,4], inhomogeneous cell currents arise within these parallel cell configurations [5], further leading to State of Charge (SoC) [6,7], Open Circuit Voltage (OCV) [8,9] and temperature gaps [6,10] within the parallel cells. Since the cell currents must be maintained in their corresponding operating window and cannot be measured for space and cost considerations, correlations of the current distribution to cell parameters and cell states are essential for an optimal and safe battery operation.

There is a variety of articles focusing on modeling [11–15], aging [16–18], safety [19–21], state estimation [22,23] and measurement [24–26] of parallel-connected cells. Mostly qualitative effects like OCV [27], SoC [13], temperature [13,26,28,29] and current differences [13] are demonstrated but quantitative relationships are missing, especially with regard to the thermal connection of the cells to neighboring cells and cooling. The cell temperature, as well as the temperature difference between parallel cells, can have a high influence on the current distribution both due to the high sensitivity of the cell resistance to the temperature [30,31] and due to the correlation of heat dissipation to the cell current [32].

Therefore, in Section 2, a test bench with a flexibly adjustable thermal connection of the individual cells to their neighboring cells and cooling is presented. Thus, both the formation of temperature gradients due to the current distribution and the impacts of design-induced temperature gradients in a battery module on the current distribution can be investigated. In Section 3, an extended dual polarization Equivalent Circuit Model (ECM) is introduced and compared to measurements. This simulation model takes into account the influences of the test bench on the current distribution and the parallel-connected cell layers within the cells. In Section 4, the influence of temperature gradients induced by inhomogeneous thermal connections of the cells to neighboring cells and cooling on the current distribution of two parallel-connected cells are investigated. At the end, a conclusion is given.

2. The Test Bench

In this section, the test bench is discussed. First, the interactions and functions of all subsystems are explained and afterwards the temperature control system is presented in more detail. Finally, the possible impacts of the test bench on the current distribution are demonstrated.

2.1. Interactions and Communication of the Subsystems

Figure 1 shows the test bench for one cell with view-optimized modifications (a), a schematic view of the temperature control system (b), a detailed view of the position and orientation of the temperature sensors (c), as well as a detailed view of the electrical connection of the cells to the cell tester highlighting the voltage and current measurement of each cell (d).

The test bench for each cell (10) consists of two Peltier Elements (11), two aluminum plates (3), two CPU coolers (1), two speed controllers (2), a measurement device (4), one micro-controller (6) and a power supply unit (7). The housing is electrically grounded and consists of aluminum profiles (5). The cell tabs are connected to the load cable via a screwed aluminum union joint (13). The cell temperature is captured by two Pt100 sensors (12) of each cell side, $T_{Cell,Center}$ and $T_{Cell,Top}$ see Figure 1c. The temperature of the adjacent aluminum plates is measured with one Pt100 sensor T_{Plate}. The cell voltage U_{cell} is measured at the cell tabs, the cell current I_{Cell} is calculated via the voltage drop at the load cable (14) of each cell, see Figure 1d. The load cables of the cells are connected to a cell tester. This setup enables one to investigate any cell topology; the number of parallel-connected cells is only limited by the size of the climate chamber. The communication and interaction of these subsystems are presented in Figure 2.

The measurements are conducted in a temperature controlled climate chamber and the battery load is controlled by a cell tester. The temperatures of the lithium-ion cells are regulated by Peltier Elements. One side of the Peltier Elements is thermally coupled to the CPU coolers to keep the temperature of this side constant at ambient temperature. In order to ensure temperature homogeneity at the cells, the Peltier Elements are embedded in aluminum plates. In addition, the plates and the cells are isolated by Polystyrene (9). The assembly of these subsystems is presented in Figure 1b. The voltages of the Peltier Elements are adjusted by speed controllers, which in turn are controlled by micro-controllers. These are regulated by a PI controller implemented in Matlab.

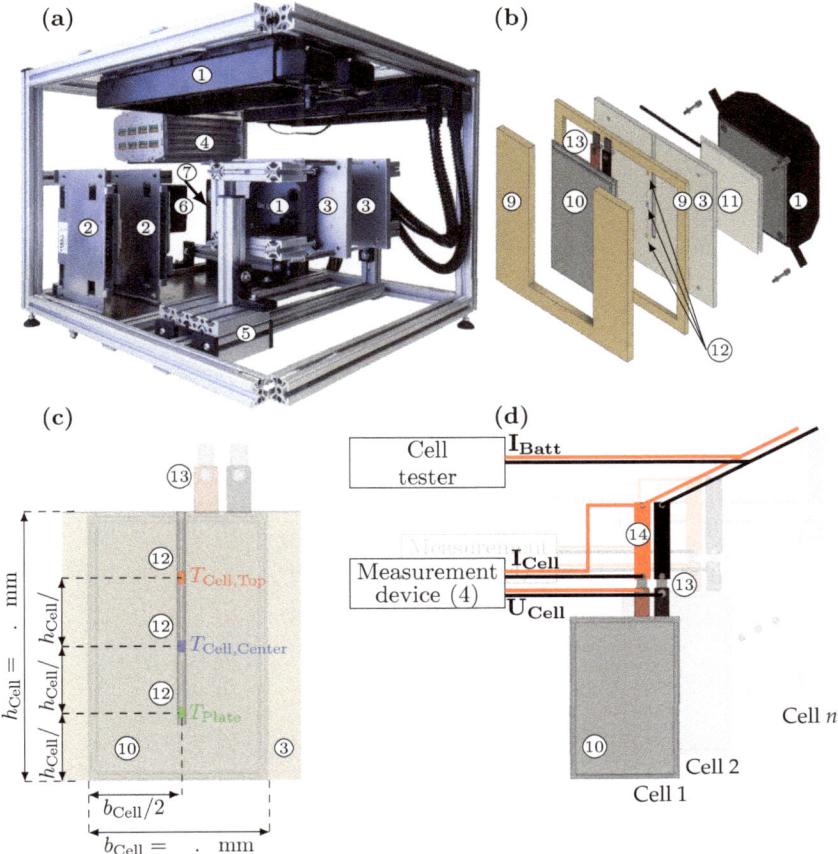

Figure 1. Setup of the test bench. View-optimized modifications of the test bench with removed cooling tubes, cables and one aluminum profile (**a**). Schematic view of the temperature control system including the positions of the Pt100 sensors (12) (**b**). Detailed view of the position and orientation of the temperature sensors (**c**) and a detailed view of the electrical connection of the cells to the cell tester (**d**).

Measurement data, including six Pt100 temperature sensors (12), three for each cell side, as well as two voltage signals, are recorded with a frequency of $f = 100$ Hz. The signals are transferred by an interface via CAN to a PC and real time processed in Matlab. The power for the speed controllers, the measurement device and CPU coolers is provided by a power supply unit.

In addition, the pressure on the cells is kept constant by a spring construction and the mobility of one aluminum plate. A fluctuation of the pressure due to the correlations of the cell's thickness to SoC [33,34] and temperature [35] as well as the continuous increase of the cell thickness caused by lithium plating and gassing [36] could otherwise have a negative effect on cell aging [37]. The manufacturers and the corresponding types of the individual components are summarized in Table 1.

The functionality of the temperature control system, temperature homogeneity of the aluminum plates and the heating rate are explained in more detail in Section 2.2.

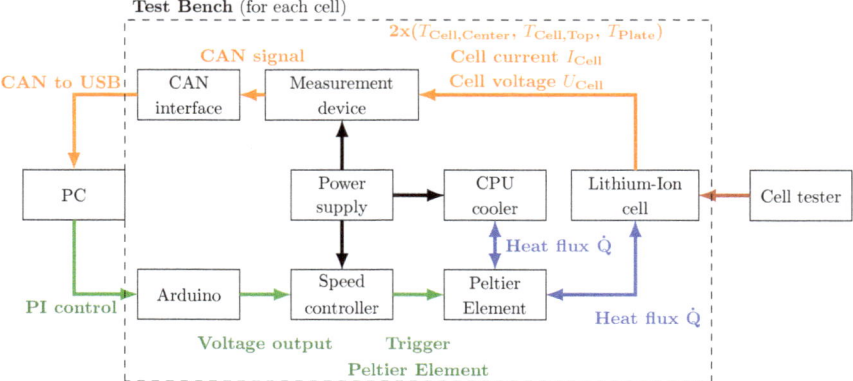

Figure 2. Schematic view of the communication and interactions of all subsystems.

Table 1. Manufacturers and corresponding types of the used components.

Designation	Manufacturer	Type
Lithium-ion cell	Kokam	SLPB776495
Peltier element	Quick Cool	QC-127-2.0-15.0M
CPU cooler	be quiet!	SL 240
Speed controller	EPH-Elektronik	DLR 24/20
Measurement device	imc CANSAS	L-HCI8
Micro-controller	Arduino	Mega 2560
Power supply unit	Thermaltake	TR2 S700
Aluminum profiles	ITEM	Profiltyp 5
CAN interface	Vector	Vn1640 A
Climate chamber	Vötsch	VC^3 4100/S
Cell tester	BasyTec	HPC

2.2. Temperature Control System

The control of the cell temperature as well as the temperature homogeneity at the cells were decisive tasks for the test bench in order to fulfill reliable and trustful measurements. In order to validate these requirements, nine temperature sensors were fixed on the cell adjacent subarea of the aluminum plate, Figure 3a, and an exemplary regulation from $T = 10\ °C$ to $T = 30\ °C$ and further to $T = 40\ °C$, Figure 3b, was conducted.

Figure 3f–h demonstrates the temperature homogeneity of the aluminum plate. The temperature distribution on the plate was determined by weighting each temperature sensor depending on their distance to the respective point. In the stationary state, the maximum temperature difference within the plate was about $\Delta T = 0.71\ °C$. The differences can be caused, on the one hand, by the smaller area of the Peltier Element compared to the aluminum plate and, on the other hand, by the energy exchange with the environment.

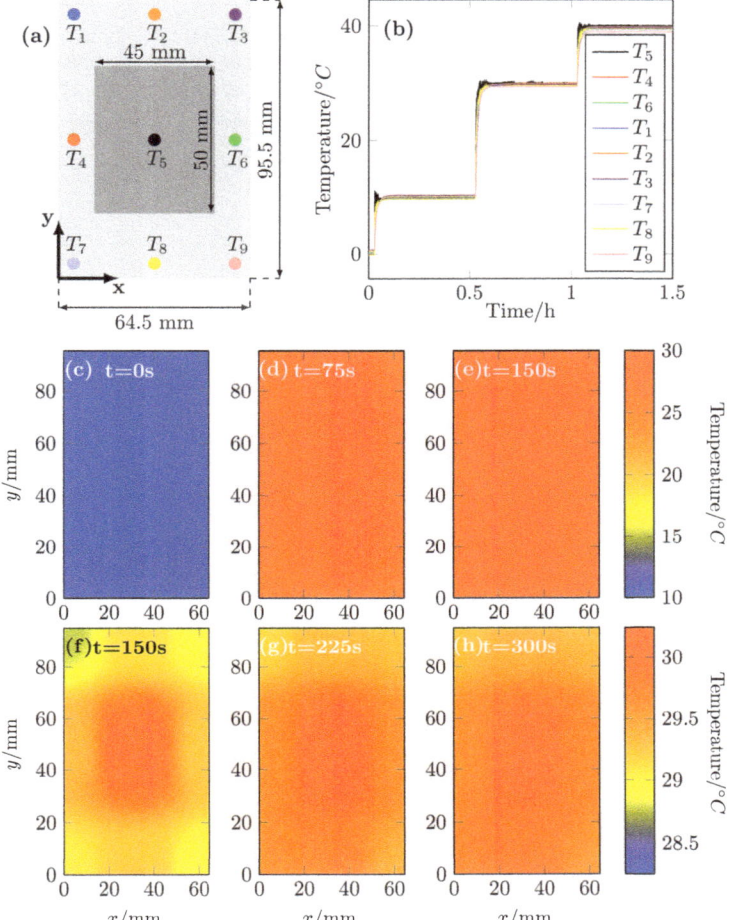

Figure 3. Investigation of the temperature homogeneity and the heating rate of the aluminum plate with the position of the Pt100 sensors and the PE (gray box) in (**a**), the time curve of the Pt100 sensors with an exemplary regulation from $T = 10\,°C$ to $T = 30\,°C$ and further to $T = 40\,°C$ (**b**), the heating rate of the aluminum plate (**c**–**e**) and the temperature homogeneity of the aluminum plate (**f**–**h**).

Another important requirement of the temperature control system is to realize a higher heating rate of the aluminum plates $\dot{T}_{\text{Plate}}$ compared to the maximum heating rate of the cells $\dot{T}_{\text{Cell}}$ caused by dissipation $\dot{Q}_{\text{Diss,max}}$, which can be estimated by

$$\dot{T}_{\text{Cell}} = \frac{1}{m_{\text{Cell}} c_p} \cdot \dot{Q}_{\text{Diss,max}}, \tag{1}$$

$$\dot{Q}_{\text{Diss, max}} = R_{\text{Cell,max}} \cdot I_{\text{Cell}}^2, \tag{2}$$

with the maximum cell resistance $R_{\text{Cell,max}}$, the cell current I_{Cell}, the cell's heat capacity $c_{p,\text{Cell}}$ and the cell mass m_{Cell}. The cell's resistance was determined by pulse tests as described in Section 3.2 and ranges between $R_{\text{Cell}} = 21.3\,\text{m}\Omega$ and $R_{\text{Cell}} = 99.44\,\text{m}\Omega$, depending on the SoC, temperature and time. With a maximum cell current of $I_{\text{Cell}} = 2C$, the required heating rate can be calculated with Equation (2) from $\dot{T}_{\text{Cell}} = 23.4\,\text{mK}\cdot\text{s}^{-1}$ to $\dot{T}_{\text{Cell}} = 114\,\text{mK}\cdot\text{s}^{-1}$. However, as the maximum currents of $I_{Cell} = 2C$ are only permitted at cell temperatures above $T_{Cell} = 10\,°C$ and the voltage limits are reached at about SoC = 8% for these high currents, the practical limit can be set to $\dot{T}_{Cell} = 74.8\,mK·s^{-1}$.

The heating rate of the aluminum plate is approximately limited to $\dot{T}_{Plate} = 250\,mK·s^{-1}$, as shown in Figure 3c–e. This enables a precise temperature adjustment of the cells and the reconstruction of temperature gradients within a battery module due to varying thermal connections of the cells to neighboring cells and cooling.

2.3. Impacts of the Test Bench on the Current Distribution

The ratios of the resistances and capacities of the parallel-connected cells have one of the main influences on the current distribution. This was shown qualitatively by measurements and simulations in [1], the quantitative influences of these parameters were mathematically proven in previous work [8,9].

For this reason, additional resistors induced by the test bench can significantly influence the current distribution. Due to the electrical connection of the cell tester and the lithium-ion cells, additional resistances arise in series with the cells, which are demonstrated in Figure 4.

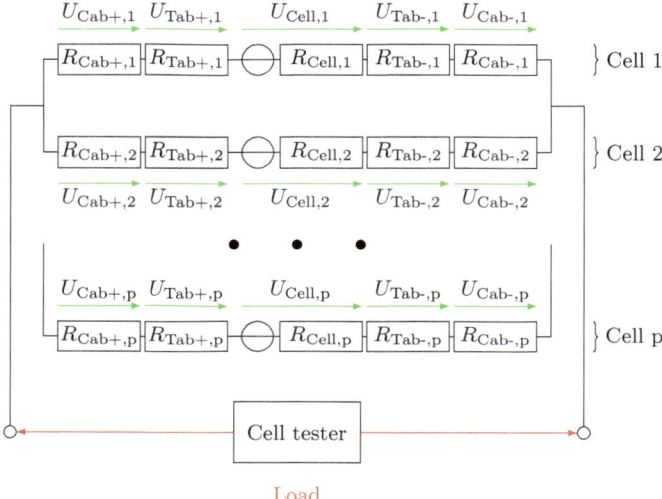

Figure 4. Possible induced resistances by the test bench influencing the current distribution of p parallel-connected cells, with the contact resistances at the positive R_{Tab+} and negative tab R_{Tab-}, the positive R_{Cab+} and negative cable resistances R_{Cab-} as well as the cell resistance R_{Cell}.

The contact resistances at the cell tabs as well as the cable resistances must on the one hand be well-known and on the other hand be kept low, at least their difference. As a result of the first Kirchhoff's law the following is valid

$$\begin{aligned}
& U_{Cab+,i} + U_{Tab+,i} + U_{Cell,i} + U_{Tab-,i} + U_{Cab-,i} \\
= & U_{Cab+,i+1} + U_{Tab+,i+1} + U_{Cell,i+1} + U_{Tab-,i+1} + U_{Cab-,i+1} \\
= & U_{Cab+,p} + U_{Tab+,p} + U_{Cell,p} + U_{Tab-,p} + U_{Cab-,p},
\end{aligned} \quad (3)$$

with the cell voltage $U_{Cell,i}$, the positive $U_{Tab+,i}$ and negative tab voltages $U_{Tab-,i}$ as well as the voltages of the positive $U_{Cab+,i}$ and negative load cable $U_{Cab-,i}$ of cell i. In order to keep these disturbing values of the test bench on the current distribution low, the following conditions must be fulfilled

$$U_{\text{Cell},1} \approx U_{\text{Cell},2} \approx U_{\text{Cell},p}, \tag{4}$$

$$U_{\text{Cab-},i} + U_{\text{Tab-},i} + U_{\text{Cab+},i} + U_{\text{Tab+},i} \ll U_{\text{Cell},i}. \tag{5}$$

The contact resistance at the tabs were examined and the cable resistances were varied in order to validate their impacts on the current distribution.

2.3.1. Contact Resistance at the Tabs

In order to keep the contact resistance at the cell tabs low, different influences on these resistances were investigated. The contact resistances were calculated by a four terminal measurement using a micro ohmmeter (MPK, 2000e). The influences of surface cleaning, torque and air insulation were researched. Therefore the tabs were treated with a non-woven abrasive cloth and then cleaned with an oxide-dissolving spray (CRC-Kontaktchemie, Kontakt 60). In addition, tests with varied contact pressure and conductive epoxy resin were conducted in order to isolate the tabs. For the investigation of the contact pressure, the resistances of twelve cathode and anode tabs were examined. To research the impacts of air seal, six of the cathode tabs were treated with conductive epoxy resin. The other six cathode tabs and the twelve anode tabs were not sealed with epoxy resin. The results are presented in Figure 5, whereby for each point the mean, minimum and maximum value is shown.

The influences of surface cleaning and contact pressure, as presented in Figure 5a,b, have already been investigated in [38,39] with the same findings. Figure 5c shows the influence of the air seal. While the contact resistances at the cathode tabs with air seal remained almost constant over the test period of 51 days, the resistances without air seal increased 2.5 times from $R_{\text{Tab},+} = 57.4\ \mu\Omega$ to $R_{\text{Tab},+} = 133.6\ \mu\Omega$. It is assumed that oxidation on the tabs will lead to this increase. A relaxation of the surface pressure during the test period could additionally have led to an increasing contact resistance. However, since the resistances at the anode tabs remained unchanged, this effect should be not that significant. The anode tabs do not require an air seal due to their nickel coating, which shows up in an unchanged resistance over the test period. Furthermore, the contact resistances at the tabs and therefore the corresponding disturbing voltage drops $U_{\text{Tab+}}$ and $U_{\text{Tab-}}$ are low compared to the cell resistance with

$$\max\left(\frac{R_{\text{Tab}}}{R_{\text{Cell}}}\right) < 1\%, \tag{6}$$

which should lead to no significant impacts on the current distribution.

2.3.2. Cable Resistance

In addition to the requirement of low resistance, the load cables were also used as a shunt to determine the cell currents, which were calculated via the voltage drop at the cables. Since this drop correlates to the cable resistance, a trade off between the measurement accuracy and the influence on the current distribution arise. Therefore different cable resistances and their impacts on the current distribution were investigated by measurements with two parallel-connected cells. The cable resistances were examined by four terminal measurements using a multimeter (Keithley, DMM7510) and are given in Table 2.

Table 2. Investigated cable resistances.

Cell	$R_{\text{Cab,Type1}}/\text{m}\Omega$	$R_{\text{Cab,Type2}}/\text{m}\Omega$	$R_{\text{Cab,Type3}}/\text{m}\Omega$
1	0.28	1.48	4.20
2	0.3	1.59	4.52

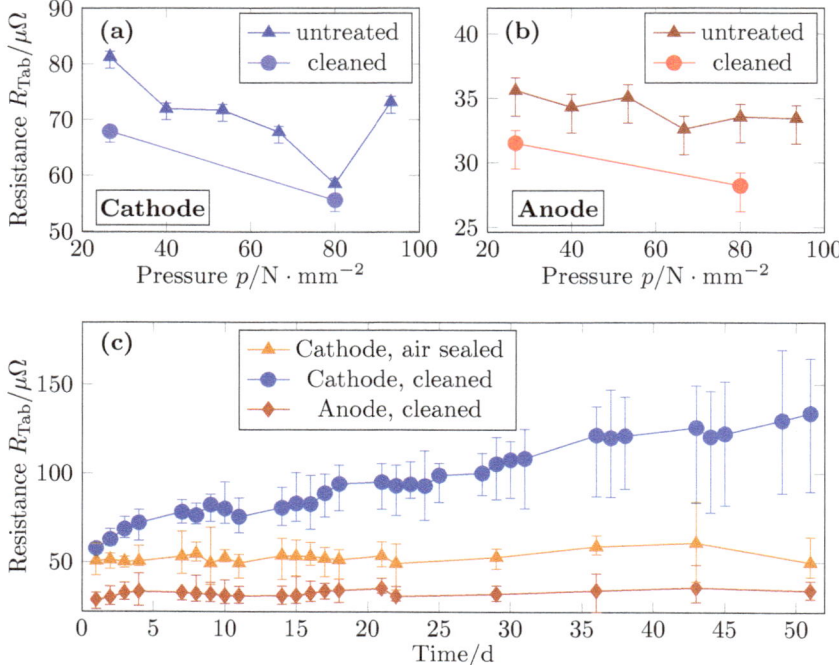

Figure 5. Impacts of tab treatment, with the influences of surface cleaning and pressure (**a,b**) as well as the effects of air sealing (**c**) on the contact resistance at the cathode (aluminum) and anode (nickel-plated copper) tab.

In relation to the cell resistance, the cable resistances of the first type $R_{Cab,Type1}$ range from 1.3% to 0.2% depending on the SoC and cell temperature and should have no significant influence on the current distribution. Type 3 $R_{Cab,Type3}$ uses the complete measuring range of the measuring device, which should lead to the highest measurement accuracy. Type 2 $R_{Cab,Type2}$ offers a trade off.

The current distribution of two parallel-connected cells with the discussed cables are presented in Figure 6. The temperature of the aluminum plates were kept constant at $T_{Plate} = 30$ °C. The cells were discharged with a current load of $I = -10.6$ A, which corresponds to a C-rate of 1C. The discharge started with a SoC = 0.95 and lasted until the cell voltage reaches the voltage limit of $U_{Cell} < 2.8$ V. After a break of 30 min, the cells were charged with a C-rate of 1C until the cell voltage reached the voltage limit of $U_{Cell} = 4.2$ V. Thereafter the cells were relaxed for 30 min.

The current distribution is qualitatively the same for the three different cables with a shift to higher current differences within the parallel-connected cells with increasing cable resistance. However, this may also be due to the higher difference between the cable resistances. The differences are most obvious at the current rest and at the end of the discharge phase. This can be caused by the correlation of the time constant to the cell's resistance, as mathematically shown in previous work [8,9].

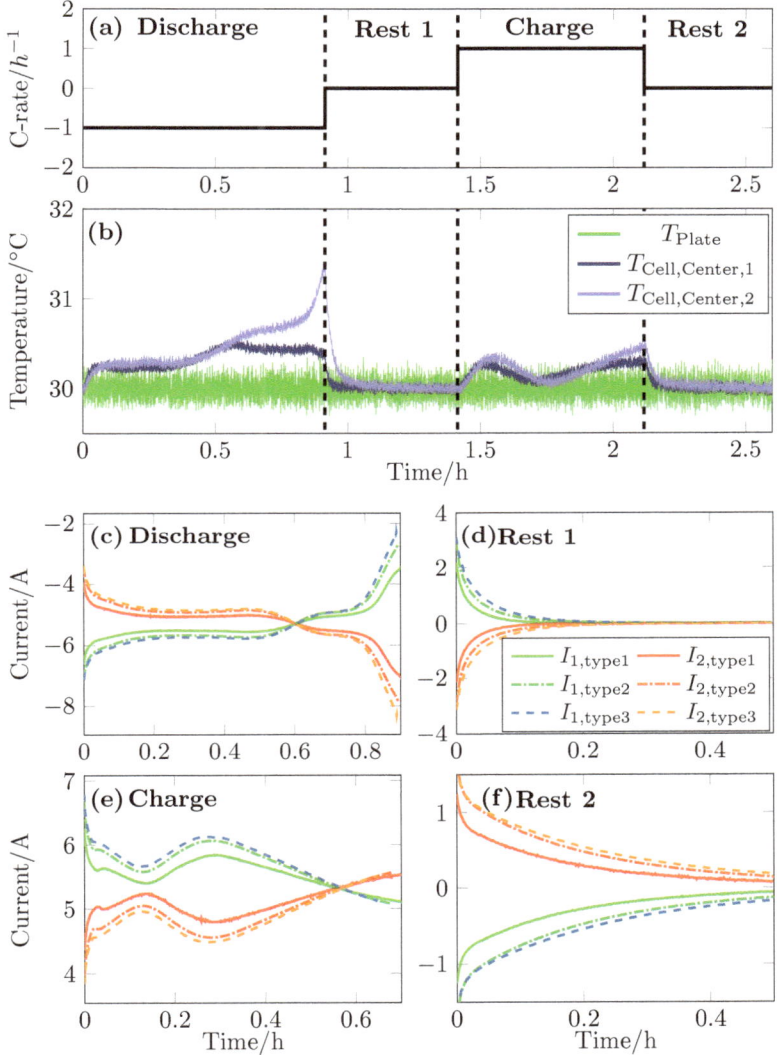

Figure 6. Influence of the cable resistances on the current distribution of two parallel-connected cells. With the current load (**a**), the cell and plate temperature (**b**) and the separated phases: Discharging (**c**), the first rest (**d**), charging (**e**) and the second rest (**f**). The relations of the cable resistances are $R_{\text{Cab,type3}} \approx 3 \cdot R_{\text{Cab,type2}} \approx 15 \cdot R_{\text{Cab,type1}}$, with the exact values in Table 2.

3. Simulation

In this section the used ECM is presented and the parametrization of the cells is discussed. At the end simulations are compared to measurements of two parallel-connected cells.

3.1. Equivalent Circuit Model

In Figure 7 the implemented dual polarization model is demonstrated. The model considers the mentioned resistances due to the connection of the cells to the cell tester and the parallel-connection of

L cell layers within the lithium-ion cell. ECMs with RC-pairs are widely used for the simulation of the current distribution within parallel-connected cells [1,2,40–43].

This ECM was connected in parallel and solved as described in previous work [8,9].

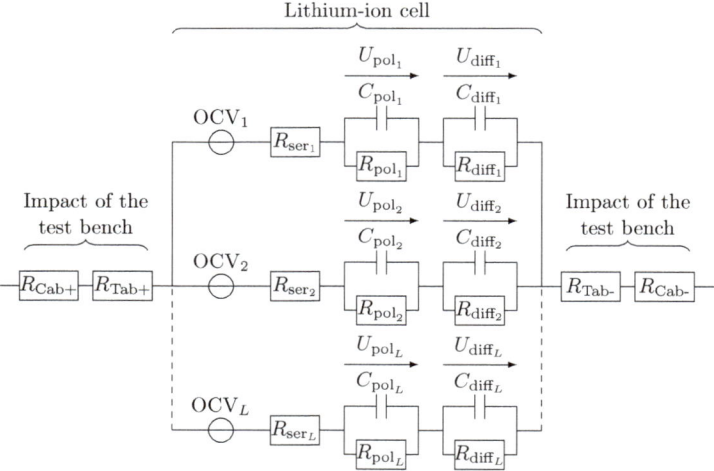

Figure 7. Extended dual polarization model—Resolution of the cell layers within the cell and consideration of the contact resistance of the tabs and cable resistances due to the connection of the cell to the cell tester.

3.2. Parametrization

The model was parameterized by Constant-Current Constant-Voltage (CCCV) measurements to estimate the cell's capacity. With a constant current of 1C until $U_{Cell} = 4.2$ V for charging, respectively $U_{Cell} = 2.8$ V for discharging. The voltage was kept constant until the cell current reached $I_{Cell} < 1/20C$. The serial resistor and the parameters of the RC-pairs were determined by 2C charge and discharge current pulses at 34 different SoCs. The cells were relaxed one hour after each pulse to estimate the OCV and guarantee same start conditions for each pulse. The measurements were repeated at different cell temperatures (The current pulses were reduced to $I_{Cell} = 0.2C$ at cell temperatures $T_{Cell} < 10\ ^\circ$C in order to keep the cells in their operating window.) $T = 0, 5, 10, 20, 30$ and $40\ ^\circ$C. The parametrization process as well as the results of the ECM parameters are presented in more detail in previous work [2].

3.3. Validation

In order to validate the discussed ECM, simulations are compared to measurements in Figure 8. Each of the used cells consists of 40 cell layers, whose resistances and capacities were randomly drawn from Gauss distributions. With the related standard deviations of the cell's resistance and the cell's capacity

$$\frac{\sigma_R}{\mu_R} = \omega_R \cdot \sqrt{L}, \tag{7}$$

$$\frac{\sigma_C}{\mu_C} = \omega_C \cdot \sqrt{L}, \tag{8}$$

with the standard deviation of the cell's resistance σ_R and the cell's capacity σ_C, the mean value of the cell's resistance μ_R and the cell's capacity μ_C as well as the number of parallel-connected cell layers L. The standard deviations of the parameters were multiplied by the square root of L to consider the statistical averaging due to the parallel cell layers. The related standard deviations ω_R and ω_C were set to $\omega_R = 1\%$ and $\omega_C = 0.5\%$, which are typical parameter distributions caused by manufacturing tolerances as found in [3,4]. The expected values μ_R and μ_C were adjusted according to the parametrization results of each cell. The load cable resistances R_{Cab} were set according to the values of Table 2. For the tab resistances R_{Tab}, the average values of the measurements in Figure 5c were used.

The results of the simulation in Figure 8 agree well with those of the measurement with an RMSD of $\zeta_{RMSD} = 0.083$ A. The highest differences appeared at the end of discharge and at the first current rest. The simulated layer currents, Figure 8a, within the cells show a similar distribution with an shift depending on the cell parameters.

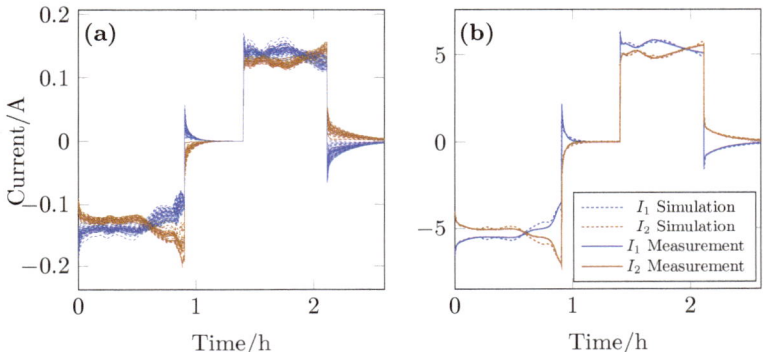

Figure 8. Comparison of measurement and simulation, with the simulated cell layer currents $I_{i,1}$ and $I_{i,2}$, $i \in \{1, \ldots, 40\}$ within cell one and two in (**a**) and the simulated cell currents as well as the measurement results of cable type 1 in (**b**).

4. Impacts of the Module Design

Due to the cell packaging, inhomogeneities in terms of cooling and thermal connection of the cells to their neighbor cells arise within a battery module. The influence of the ambient temperature depends on the cell position in the module, whereby the cells at the edge of the module are mainly affected. This leads to inhomogeneous cell cooling resulting in temperature gradients. Existing tension mats, which thermally isolate the cells, can further intensify arising temperature gradients. Possible scenarios are presented in Figure 9.

In order to investigate the influence of the cell position and thermal connection of cells, the tension mats were modeled as heat impermeable and the module housing was assumed to be isotherm with a constant temperature at ambient temperature. For each scenario, which considers the effects of the border cell, two measurements were conducted to switch the border position of the two parallel-connected cells. The reference represents conditions at a constant cell temperature, which rather corresponds to most publications.

The current and temperature distribution of these scenarios for two parallel-connected cells are displayed in Figure 10. The cells were fully discharged with $I_{Batt} = 1C$ until the cell pack reached the lower voltage limit of $U_{Batt} = 2.8$ V. Thereafter, the cells were further discharged at constant voltage of $U_{Batt} = 2.8$ V until the battery current decreased to $I_{Batt} = 1/20$. After a relaxation of $T_{Break} = 1$ h, the parallel-connected cells were charged with $I_{Batt} = 1C$ until the upper voltage limit of $U_{Batt} = 4.2$ V was reached. The ambient temperature and the initial temperature of the aluminum plates were set to $T = 20\,°C$.

The scenarios show significant differences of the current and temperature distribution of the parallel-connected cells. In the following the scenarios will be separately discussed.

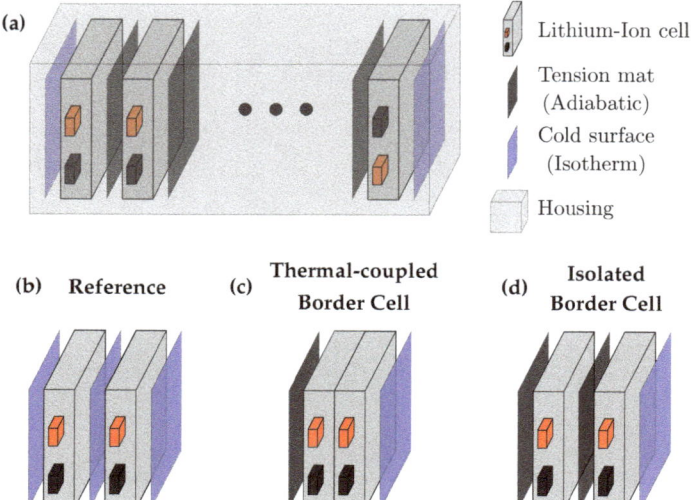

Figure 9. Impacts of the cell position and isolating tension mats on temperature gradients within a battery module (**a**). Considered scenarios: Reference (**b**), thermal-coupled border cell (**c**) and isolated border cell (**d**).

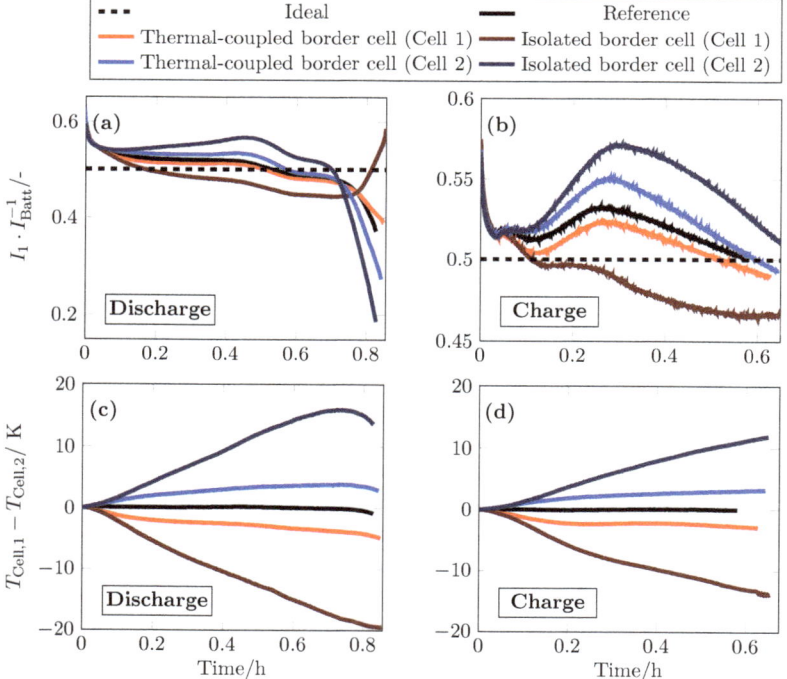

Figure 10. Measurements simulating different cell positions in a battery and thermal connections of cells to their neighbor cells of two parallel-connected cells. With the current (**a**) and temperature distribution (**c**) at discharge with $I_{\text{Batt}} = 1C$ as well as the current (**b**) and temperature distribution (**d**) at charge with $I_{\text{Batt}} = 1C$.

Cell 2—Border Cell

With cell two as border cell, temperature gaps of $\Delta T = 3.8\,°C$ for the thermal-coupled and of $\Delta T = 15.8\,°C$ for the thermal isolated cells raised at discharge. This caused higher current differences and delayed crossing points of the cell currents and resulted in increasing SoC gaps within the parallel cells. For discharging this significantly influenced the current distribution after the crossing point. The OCV bending at low SoC led to a rapid voltage drop of cell one, which resulted in a current peak of cell two. This effect is more distinct with increasing SoC difference, which shows up in an increasing current gap from $I_2 = 0.73 \cdot I_{Batt}$ to $I_2 = 0.81 \cdot I_{Batt}$ comparing the thermal-coupled and the isolated scenario. For charging similar effects can be seen. As the current load starts at the point of the highest OCV bending and therefore no distinct SoC gaps have been raised, the rapid current decrease is not seen for charging.

Cell 1—Border Cell

In this scenario the temperature difference showed the same affects on the current distribution, but considering the current of cell one in opposite direction. The temperature difference led to a lower current of cell one and an earlier starting of the crossing point, even compared to the reference measurement. The temperature gap in the thermal-coupled scenario affect more balanced cell currents compared to the reference. This led also to a lower SoC gap, which in turn reduced the current peak of cell two to $I_2 = 0.61 \cdot I_{Batt}$. The isolation further decreases the current of cell one I_1, which caused a higher discharge and charge rate of cell two compared to cell one.

5. Conclusions

In this article a test bench was presented, which enables the individual temperature control of each cell connected in parallel. This allows to reconstruct arising temperature gradients in a battery module due and to investigate their impacts on the dynamic of the current distribution. The aluminum plates adjacent to the cells can be heated and cooled with a rate of about $\dot{T}_{Plate} = 250\,mK \cdot s^{-1}$, which exceeds the maximum heating rate of the cells due to dissipation with $\dot{T}_{Cell} = 114\,mK \cdot s^{-1}$. The influence of the test bench on the current distribution caused by induced additional resistors was determined and minimized. The contact resistance at the cathode tab was reduced from $R_{Tab+} = 81.18\,\mu\Omega$ to $R_{Tab+} = 55.15\,\mu\Omega$ by treating with a non-woven abrasive cloth, cleaning with an oxide-dissolving spray and increasing the pressure from $p = 27\,N \cdot mm^{-2}$ to $p = 93\,N \cdot mm^{-2}$. In addition, an increase of the contact resistance during the test period is prevented by air seal of the contact. Without air sealing, the resistance increased from $R_{Tab+} = 57.54\,\mu\Omega$ to $R_{Tab+} = 133.57\,\mu\Omega$ within 51 days at room temperature. The contact resistance at the anode tab was reduced by the same treatments from $R_{Tab-} = 35.59\,\mu\Omega$ to $R_{Tab-} = 28.2\,\mu\Omega$. Due to their nickel coating an air seal was not necessary.

Measurements of two parallel-connected cells with load cable resistances of $R_{Cab+} = 0.3\,m\Omega$, $R_{Cab+} = 1.6\,m\Omega$ and $R_{Cab+} = 4.35\,m\Omega$ showed qualitatively the same dynamic of the current distribution with decreasing current differences within the parallel-connected cells with decreasing cable resistance. An ECM considering the current distribution within the cells as well as the impacts of the induced resistances by the test bench was introduced, parameterized and compared to measurements. The model fitted well to measurements with an RMSD of $\zeta_{RMSD} = 0.083\,A$. Measurements simulating different cell positions in a battery and thermal connections of cells to their neighbor cells were conducted. The consideration of a cell at the module edge showed increasing temperature differences of $\Delta T = 3.8\,°C$ for thermal-coupled cells and of $\Delta T = 15.8$ for thermal isolated cells. This temperature difference further increased the initial cell parameter difference and led to higher current and SoC gaps within the parallel-connected cells. The crossing point of the cell current was delayed with increasing ΔT, which in turn caused an increasing current peak of cell two from $I_2 = 0.73 \cdot I_{Batt}$ to $I_2 = 0.81 \cdot I_{Batt}$ comparing the thermal-coupled and isolated scenarios.

Author Contributions: A.F., T.M., T.S. and R.L. conceived and designed the test bench. A.F., T.M. and T.S. conducted the experiments and analyzed the data. A.F. designed the simulations and wrote the article. K.P.B. and T.S. contributed to the manuscript design and revised the article. All authors have read and agreed to the published version of the manuscript.

Funding: This research received no external funding.

Acknowledgments: The authors want to thank Björn Mulder for reviewing the article.

Conflicts of Interest: The authors declare no conflict of interest.

Abbreviations

The following abbreviations are used in this manuscript:

RMSD   Root Mean Square Deviation
ECM   Equivalent Circuit Model
SoC   State of Charge
OCV   Open Circuit Voltage
CPU   Central Processing Unit
CAN   Controller Area Network

References

1. Brand, M.J.; Hofmann, M.H.; Steinhardt, M.; Schuster, S.F. Current distribution within parallel-connected battery cells. *J. Power Sources* **2016**, *334*, 202–212. [CrossRef]
2. Fill, A.; Birke, K.P. Impacts of cell topology, parameter distributions and current profile on the usable power and energy of lithium-ion batteries. In Proceedings of the 2019 International Conference on Smart Energy Systems and Technologies (SEST), Porto, Portugal, 9–11 September 2019; pp. 1–6.
3. Rumpf, K.; Naumann, M.; Jossen, A. Experimental investigation of parametric cell-to-cell variation and correlation based on 1100 commercial lithium-ion cells. *J. Energy Storage* **2017**, *14*, 224–243. [CrossRef]
4. Devie, A.; Baure, G.; Dubarry, M. Intrinsic Variability in the Degradation of a Batch of Commercial 18650 Lithium-Ion Cells. *Energies* **2018**, *11*, 1031. [CrossRef]
5. Pastor-Fernandez, C.; Bruen, T.; Widanage, W.; Gama-Valdez, M.; Marco, J. A Study of Cell-to-Cell Interactions and Degradation in Parallel Strings:Implications for the Battery managment System. *J. Power Sources* **2016**, *329*, 574–585. [CrossRef]
6. Baumann, M.; Wildfeuer, L.; Rohr, S.; Lienkamp, M. Parameter variations within Li-Ion battery packs—Theoretical investigations and experimental quantification. *J. Energy Storage* **2018**, *18*, 295–307. [CrossRef]
7. Hofmann, M.H.; Czyrka, K.; Brand, M.J.; Steinhardt, M.; Noel, A.; Spingler, F.B.; Jossen, A. Dynamics of current distribution within battery cells connected in parallel. *J. Energy Storage* **2018**, *20*, 120–133. [CrossRef]
8. Fill, A.; Koch, S.; Pott, A.; Birke, K.P. Current distribution of parallel-connected cells in dependence of cell resistance, capacity, energy density and number of parallel cells. *J. Power Sources* **2018**, *407*, 147–152. [CrossRef]
9. Fill, A.; Koch, S.; Birke, K.P. Analytical model of the current distribution of parallel-connected battery cells and strings. *J. Energy Storage* **2019**, *23*, 37–43. [CrossRef]
10. Yang, N.; Zhang, X.; Shang, B.; Lia, G. Unbalanced discharging and aging due to temperature differences among the cells in a lithium-ion battery pack with parallel combination. *J. Energy Storage* **2016**, *306*, 733–741. [CrossRef]
11. Yang, N.; Zhang, X.; Shang, B.; Lia, G. A Novel Dynamic Performance Analysis and Evaluation Model of Series-Parallel Connected Battery Pack for Electric Vehicles. *IEEE Access* **2019**, *7*, 14256–14265.
12. Liu, X.; Ai, W.; Marlow, M.N.; Patel, Y.; Wu, B. The effect of cell-to-cell variations and thermal gradients on the performance and degradation of lithium-ion battery packs. *Appl. Energy* **2019**, *248*, 489–499. [CrossRef]
13. Hosseinzadeh, E.; Marco, J.; Jennings, P. Combined electrical and electrochemical-thermal model of parallel connected large format pouch cells. *J. Energy Storage* **2019**, *22*, 194–207. [CrossRef]
14. Dubarry, M.; Pastor-Fernández, C.; Baure, G.; Yu, T.F.; Widanage, W.D.; Marco, J. Battery energy storage system modeling: Investigation of intrinsic cell to cell variations. *J. Energy Storage* **2019**, *23*, 19–28. [CrossRef]

15. Kakimoto, N.; Goto, K. Capacity-Fading Model of Lithium-Ion Battery Applicable to Multicell Storage Systems. *IEEE Trans. Sustain. Energy* **2016**, *7*, 108–117. [CrossRef]
16. Gogoana, R.; Pinson, M.B.; Bazant, M.Z.; Sarma, S.E. Internal resistance matching for parallel-connected lithium-ion cells and impacts on battery pack cycle life. *J. Power Sources* **2014**, *252*, 8–13. [CrossRef]
17. Shi, W.; Hu, X.; Chao, J.; Jiang, J.; Zhang, Y.; Yip, T. Effects of imbalanced currents on large-format $LiFePO_4$/graphite batteries systems connected in parallel. *J. Power Sources* **2016**, *313*, 198–204. [CrossRef]
18. Wang, X.; Wang, Z.; Wang, L.; Wang, Z.; Guo, H. Dependency analysis and degradation process-dependent modeling of lithium-ion battery packs. *J. Power Sources* **2019**, *414*, 318–326. [CrossRef]
19. Fill, A.; Koch, S.; Birke, K.P. Algorithm for the detection of a single cell contact loss within parallel-connected cells based on continuous resistance ratio estimation. *J. Energy Storage* **2020**, *27*, 101049. [CrossRef]
20. Wang, L.; Cheng, Y.; Zhao, X. A $LiFePO_4$ battery pack capacity estimation approach considering in-parallel cell safety in electric vehicles. *Appl. Energy* **2015**, *142*, 293–302. [CrossRef]
21. Koch, S.; Fill, A.; Birke, K.P. Discharge by Short Circuit Currents of Parallel-Connected Lithium-Ion Cells in Thermal Propagation. *Batteries* **2019**, *5*, 18. [CrossRef]
22. Barai, A.; Ashwin, T.; Iraklis, C.; McGordon, A.; Jennings, P. Scale-up of lithium-ion battery model parameters from cell level to module level—Identification of current issues. *Energy Procedia* **2017**, *138*, 223–228. [CrossRef]
23. Li, J.; Greye, B.; Buchholz, M.; Danzer, M.A. Interval method for an efficient state of charge and capacity estimation of multicell batteries. *J. Energy Storage* **2017**, *13*, 1–9. [CrossRef]
24. Hunt, I.; Zhang, T.; Patel, Y.; Marinescu, M.; Purkayastha, R.; Kovacik, P.; Walus, S.; Swiatek, A.; Offer, G. The Effect of Current Inhomogeneity on the Performance and Degradation of Li-S Batteries. *J. Electrochem. Soc.* **2018**, *165*, 6073–6080. [CrossRef]
25. Gong, X.; Xoing, R.; Mi, C. Study of the Characteristics of Battery Packs in Electric Vehicles with Parallel-Connected Lithium-Ion Battery Cells. *IEEE Trans. Ind. Appl.* **2014**, *51*, 1872–1879. [CrossRef]
26. Klein, M.P.; Park, J.W. Current Distribution Measurements in Parallel-Connected Lithium-Ion Cylindrical Cells under Non-Uniform Temperature Conditions. *J. Electrochem. Soc.* **2017**, *164*, 1893–1906. [CrossRef]
27. Dubarry, M.; Devie, A.; Liaw, B.Y. Cell-balancing currents in parallel strings of a battery system. *J. Power Sources* **2016**, *321*, 36–46. [CrossRef]
28. Zhao, C.; Cao, W.; Dong, T.; Jiang, F. Thermal behavior study of discharging/charging cylindrical lithium-ion battery module cooled by channeled liquid flow. *Int. J. Heat Mass Transf.* **2018**, *120*, 751–762. [CrossRef]
29. LeBel, F.A.; Wilke, S.; Schweitzer.; Roux, M.A.; Al-Hallaj, S.; Trovao, J. A Lithium-Ion Battery Electro-Thermal Model of Parallellized Cells. In Proceedings of the 2016 IEEE 84th Vehicular Technology Conference (VTC-Fall), Montreal, QC, Canada, 18–21 September 2016.
30. Zhang, C.; Li, K.; Song, J.D.S. Improved Realtime State-of-Charge Estimation of $LiFePO_4$ Battery Based on a Novel Thermoelectric Model. *IEEE Trans. Ind. Electron.* **2017**, *64*, 654–663. [CrossRef]
31. Widanage, W.; Barai, A.; Chouchelamane, G.; Uddin, K.; McGordon, A.; Marco, J.; Jennings, P. Desing and use of multisine signals for Li ion battery equivalent circuit modeling. Part 2: Model Estimation. *J. Power Sources* **2016**, *324*, 61–69.
32. Chacko, S.; Chung, Y.M. Thermal modelling of Li-ion polymer battery for electric vehicle drive cycles. *J. Power Sources* **2012**, *213*, 296–303. [CrossRef]
33. Oh, K.Y.; Siegel, J.B.; Secondo, L.; Kim, S.U.; Samad, N.A.; Qin, J.; Anderson, D.; Garikipati, K.; Knobloch, A.; Epureanu, B.I.; et al. Rate dependence of swelling in lithium-ion cells. *J. Power Sources* **2014**, *267*, 197–202. [CrossRef]
34. Grimsmann, F.; Brauchle, F.; Gerbert, T.; Gruhle, A.; Knipper, M.; Parisi, J. Hysteresis and current dependence of the thickness change of lithium-ion cells with graphite anode. *J. Energy Storage* **2017**, *12*, 132–137. [CrossRef]
35. Oh, K.Y.; Epureanu, B.I. A novel thermal swelling model for a rechargeable lithium-ion battery cell. *J. Power Sources* **2016**, *303*, 86–96. [CrossRef]
36. Grimsmann, F.; Brauchle, F.; Gerbert, T.; Gruhle, A.; Knipper, M.; Parisi, J.; Knipper, M. Impact of different aging mechanisms on the thickness change and the quick-charge capability of lithium-ion cells. *J. Energy Storage* **2017**, *14*, 158–162. [CrossRef]
37. Cannarella, J.; Arnold, C.B. Stress evolution and capacity fade in constrained lithium-ion pouch cells. *J. Power Sources* **2014**, *245*, 745–751. [CrossRef]

38. Brand, M.J.; Berg, P.; Kolp, E.I.; Bachb, T.; Schmidt, P.; Jossen, A. Detachable electrical connection of battery cells by press contacts. *J. Energy Storage* **2016**, *8*, 69–77. [CrossRef]
39. Bolsinger, C.; Zorn, M.; Birke, K.P. Electrical contact resistance measurements of clamped battery cell connectors for cylindrical 18650 battery cells. *J. Energy Storage* **2017**, *12*, 29–36. [CrossRef]
40. Bruen, T.; James, M. Modelling and experimental evaluation of parallel connected lithium ion cells for an electric vehicle battery system. *J. Power Sources* **2016**, *310*, 91–101. [CrossRef]
41. Zhang, J.; Ci, S.; Sharif, H.; Alahmad, M. Modeling Discharge Behavior of Multicell Battery. *IEEE Trans. Energy Convers.* **2010**, *25*, 1133–1141. [CrossRef]
42. Grün, T.; Stella, K.; Wollersheim, O. Influence of circuit design on load distribution and performance of parallel-connected Lithium ion cells for photovoltaic home storage systems. *J. Energy Storage* **2018**, *17*, 367–382. [CrossRef]
43. Cordoba-Arenas, A.; Onori, S.; Rizzoni, G. A control-oriented lithium-ion battery pack model for plug-in hybrid electric vehicle cycle-life studies and system design with consideration of health management. *J. Power Sources* **2015**, *279*, 791–808. [CrossRef]

© 2019 by the authors. Licensee MDPI, Basel, Switzerland. This article is an open access article distributed under the terms and conditions of the Creative Commons Attribution (CC BY) license (http://creativecommons.org/licenses/by/4.0/).

Article

Development of a Polymeric Arrayed Waveguide Grating Interrogator for Fast and Precise Lithium-Ion Battery Status Monitoring

Jan Meyer [1,*], Antonio Nedjalkov [1,2], Elke Pichler [2], Christian Kelb [1] and Wolfgang Schade [1,2]

[1] Department for Fiber Optical Sensor Systems, Fraunhofer Heinrich Hertz Institute, 38640 Goslar, Germany; antonio.nedjalkov@hhi.fraunhofer.de (A.N.); christian.kelb@hhi.fraunhofer.de (C.K.); wolfgang.schade@hhi.fraunhofer.de (W.S.)

[2] Institute of Energy Research and Physical Technologies, Clausthal University of Technology, 38640 Goslar, Germany; elke.pichler@tu-clausthal.de

* Correspondence: jan.meyer@hhi.fraunhofer.de

Received: 10 September 2019; Accepted: 14 October 2019; Published: 18 October 2019

Abstract: We present the manufacturing and utilization of an all-polymer arrayed waveguide grating (AWG) interacting with a fiber Bragg grating (FBG) for battery status monitoring on the example of a 40 Ah lithium-ion battery. The AWG is the main component of a novel low-cost approach for an optical interrogation unit to track the FBG peak wavelength by means of intensity changes monitored by a CMOS linear image sensor, read out by a Teensy 3.2 microcontroller. The AWG was manufactured using laser direct lithography as an all-polymer-system, whereas the FBG was produced by point-by-point femtosecond laser writing. Using this system, we continuously monitored the strain variation of a battery cell during low rate charge and discharge cycles over one month under constant climate conditions and compared the results to parallel readings of an optical spectrum analyzer with special attention to the influence of the relative air humidity. We found our low-cost interrogation unit is capable of precisely and reliably capturing the typical strain variation of a high energy pouch cell during cycling with a resolution of 1 pm and shows a humidity sensitivity of −12.8 pm per %RH.

Keywords: arrayed waveguide grating (AWG); CMOS sensor; direct laser lithography; fiber Bragg grating (FBG); lithium-ion battery

1. Introduction

Lithium-ion batteries have become the foundation of a wide variety of applications depending on electrochemical energy storage [1], starting with small single cells in billions of mobile phones [2] to a vigorously growing amount of electric vehicles [3] and large battery storage power stations [4], although the latter is oftentimes viewed critically under economic aspects [5]. Regardless of the size or scope, all lithium-ion batteries have in common the need to be meticulously monitored in order to ensure safe and durable operation. This type of electrochemical energy storage only has a small range of tolerated operational states [6] and tend toward exothermic reactions, usually referred to as thermal runaway (TR) [7], if the limits for safe operation are violated.

If voltage or temperature specifications are violated, the resulting fire can quickly propagate through densely packed battery systems, causing the TR of neighbored cells and ultimately of the whole system. A number of accidents have already attracted public attention in the past [8] and it is safe to say, that, with increasing number of electric cars and high-powered stationary applications this attention will increase.

Additional to ensuring operational safety, adequate status monitoring enables the user to decide whether a lithium-ion battery may be used any further [9]; for example, a used traction battery with an insufficient capacity may still operate fine in low power stationary applications. Extended monitoring of lithium-ion batteries therefore can affect the economic treatment of used batteries [10] as e.g., either waste or valued energy storage systems for stationary applications.

State-of-the-art monitoring systems for lithium-ion batteries consist almost exclusively of electronic battery management systems (BMS) that obtain their information from electrical sensors [11]. The most frequently obtained measurement quantities are voltage, temperature and current. While voltage is usually acquired for each cell, current and temperature are often measured for modules and/or whole battery packs. A wide variety of methods for determining the state of charge (SOC) is currently utilized ranging from simple charge counting over model-based observers to neural networks and fuzzy logic algorithms. All methods yield a SOC accuracy equal or below 6% although diminishing capacity due to wear on the battery is not modeled in SOC estimation. State of health (SOH) estimation is therefore carried out in parallel fashion [12,13], using e.g., different battery models and the results are fed into the mostly adaptive SOC algorithms.

However, most applied methods still lack on accuracy and a safety margin for allowed battery SOC and SOH is usually applied to account for uncertainties and avoid financial damages e.g., due to warranty claims.

Additional information about the SOH of a lithium-ion battery would therefore be desirable and can for instance be obtained by strain sensors due to increasing distension over the lifetime, causing e.g., strain on clamped battery packs [14]. Electric strain sensors however are only partially suited for the task, since precise measurement is only possible using bulky 3- or 4-wire connections and strong electric currents inside the battery system can impact the measurement.

One well-known optical method for strain and temperature measurement is the application of fiber Bragg gratings (FBG), which combine many advantages [15]. They are small and lightweight, and a large number of sensor elements can be cost-effectively integrated in a single glass fiber, thus reducing cabling complexity. They are mostly insensitive to electric and magnetic fields and able to withstand high temperatures of several hundred degrees Celsius [16,17].

FBG sensors have already been used for status monitoring of lithium-ion batteries by different research groups [18], since they are capable of delivering information beyond conventional BMS. They can serve as simple temperature sensors [19,20], outside and even inside of a cell [21,22], or as functionalized sensor elements that are sensitive to chemical substances and their optical properties [23,24]. Additionally, due to the small fiber diameter of 125 µm, FBG sensors can be integrated in many areas that are usually inaccessible for electric sensors e.g., in a battery system consisting of many densely stacked cells. FBG, applied as external strain sensors for monitoring lithium-ion cells with a flexible casing (also referred to as pouch cells), have been demonstrated several times already [25–27]. This observable strain variation during operation is of great interest for an improved state estimation because none of the aforementioned methods used in practice are capable of delivering information about the mechanical behavior of the cells, e.g., swelling due to aging effects.

For the readout of FBG sensors, several different interrogators exist [28]. Interferometric methods [29,30], tunable-filter-based interrogation methods [31] or methods with conversion from the wavelength domain to the time domain [32] are described in literature. All mentioned methods however are realized in discrete, bulky optical setups, built and adjusted by hand, even in the case of commercial devices and not robust enough, e.g., against vibrations, for reliable utilization in automotive applications.

Arrayed waveguide gratings (AWG), commonly used as (de)-multiplexing devices in telecommunications [33], are also well known as FBG interrogators for many years and their advantages, such as high number of output channels, wide bandwidth, precise wavelength detection, high-speed capability and low cost, have been reported [34]. Additionally, their compact, integrated design leads to a certain robustness that discrete systems cannot offer.

Most AWG are made of silicon and operate at wavelengths around 1550 nm [35]. In special applications—as it is the case with FBG-based battery monitoring at least today—the costs of manufacturing process setup as well as the cost-per-unit for silicon AWG is so high, that we could find no scientific or commercial implementation of AWG-based readout for fiber optic battery sensors despite the huge advantages. We therefore manufactured all-polymer AWG that are working at near IR center wavelength (850 nm), reducing manufacturing complexity, post-processing effort and resulting per-unit cost dramatically, as described in [36]. A further advantage of the unit working in the 850 nm region lies in lower cost of peripheral components like light sources and detectors.

Here, we present to our knowledge the first interrogation system based on an all-polymer AWG that is read-out by a CMOS linear image sensor and utilized as a readout-device for fiber optic battery sensors.

2. Sensor and System Design

In the following subsections, the key components of the novel interrogation system are presented consecutively. Although the mathematical modeling, simulation and general fabrication steps for an all-polymeric AWG have been recently presented [36] and are not part of this work, the design and manufacturing are described as differences exist.

2.1. Arrayed Waveguide Grating Fabrication

The AWG is the central sensor component of the experiments carried out in this work. The main goal, therefore, is to realize a reproducible and long-term stable status monitoring with an all-polymer AWG, which also achieves the required measurement accuracy for this purpose. Since initially only one optical measuring point is to be evaluated, in the first development step, an AWG with one input and three output channels is utilized. The latter have their intensity maximum at the wavelengths 850.6, 851.6 and 852.6 nm, respectively.

Before production, an optical simulation is performed with a commercially available software program (Epipprop, Photon Design), taking into account the refractive index data and attenuation losses of the materials used as well as the resolution capability of the production machine applied. In order to minimize possible stray light effects, input and output waveguides should have an angular offset of 90°. The simulation result represents a compromise of high transmission efficiency, high channel sensitivity and low modal dispersion. The finalized design consists of 50 single arrayed waveguides with a grating order of 40 and a radius of 1.5 mm. At the input and output of this arrayed waveguides, the two free propagation zones are arranged, which each are assigned a length of 4.0 mm by the simulation. For the best possible light wave guidance at the relevant wavelength, a layer height of the entire polymeric structure of 3.2 µm is determined. The AWG is calculated with a waveguide width of 5.2 µm causing all waveguides to guide a second mode over the fundamental one. Due to the manufacturing process, a minimum distance of 1.0 µm must be maintained between the individual waveguides, which correlates with the minimum pattern size of the applied sensor structuring direct laser lithography machine (µPG 101, Heidelberg Instruments). This resolution also dictates the non-single-mode waveguides—however, the negative effects of this are satisfactorily compensated by the relatively low grating order.

In the next step, a graphic production template for the fabrication is created from the simulation result. As substrate material Cyclo-olefin polymer (COP, Zeonex® flexible foil, Microfluidic ChipShop, Jena, Germany) with a thickness of 188 µm is used. Its surface is treated with oxygen plasma for 1 min (Plasma Prep, Gala Instruments, Bad Schwalbach, Germany) to effect favorable adhesion with the photoconductive polymer, which consists of an inorganic–organic copolymer system (EpoCore/EpoClad, microresist technology). At first, EpoClad as the lower cladding with a refractive index of 1.5708 is spin-coated to a height of 2 µm. This layer is then pre-baked for 5 min at 120 °C, subsequently cured by 365 nm UV flood exposure and lastly hard-baked for 60 min at 120 °C. After another surface plasma treatment for 1 min, EpoCore as the light-guiding patterned material with a refractive index of 1.5836 is spin-coated to a height of 3.2 µm. According to the graphic template, the sensor element structure is created by laser direct patterning with the lithography machine at a wavelength of 375 nm and a power of 6 mW. This is followed by a post-bake phase for 5 min at 90 °C, before the non-cured areas are removed with the developer (mrDev600, microresist technology). The remaining cured AWG structure is subsequently hard-baked for 60 min at 120 °C. In the last step, EpoClad as upper cladding is again applied with a height of 20 µm by spin coating, hereafter pre-baked, UV flood exposed and finally post-baked. In Figure 1, a total-view microscope image of the AWG structure and associated height profile measurements are shown.

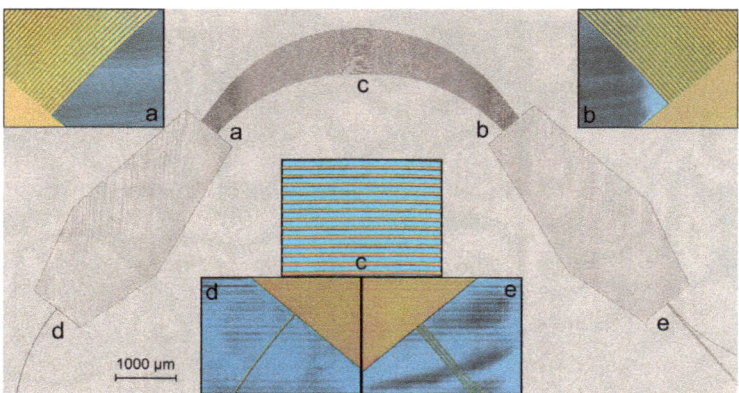

Figure 1. Laser scanning microscope images of the arrayed waveguide grating (AWG) manufactured with the described parameters. Between the end of the first free propagation zone (a) and the beginning of the second free propagation zone (b), the 50 arrayed waveguides are arranged (c). At the entrance to the first free propagation zone is the input waveguide (d) and at the exit of the second free propagation zone are the three output waveguides (e). The height of the entire structure is homogeneously 3.2 µm.

In order to have low insertion losses, the input waveguide is vertically cut and then polished. The output waveguides are treated in the same way. In the next section, the integration of the polymer AWG into the interrogator is described.

2.2. Interrogator Design

The previously presented AWG is the main component of the interrogator. To eliminate influences of the temperature on the optical properties of the AWG, it is fixed to an isothermal plate with a set temperature of 25 °C and regulated by a Peltier controller (MTD415L, Thorlabs). After polishing, a single-mode fiber is glued to the AWG polymeric input waveguide. At the other end of the fiber, a FC/APC connector is attached and is connected to a fiber port. The output waveguides face directly to a vertical CMOS linear image Sensor (iC-LFH1024, iC-Haus, Bodenheim, Germany). The sensor has a total of 1024 pixels, each with 600 µm height and 12.7 µm width. The output with its intensity maximum for a wavelength of 852.6 nm is positioned at pixel 407, the output with its intensity peak at a wavelength of 851.6 nm is positioned at pixel 486, and the third output with its intensity maximum at a wavelength of 850.6 nm is positioned at pixel 565, respectively. The descending order of wavelength at an increasing order of pixel number is simply because the CMOS linear image sensor is fixed up-side-down to the isothermal plate for easier connectorization. The CMOS linear image sensor is controlled and read-out by a microcontroller (Teensy 3.2, PJRC), that transmits its data via USB to a primary computer where they are processed further. The described components are placed in a lightproof aluminum casing. A schematic overview is shown in Figure 2.

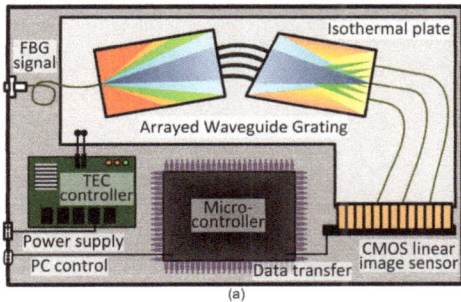

Figure 2. Schematic (**a**) and photographic (**b**) overview of the AWG interrogator. The polymeric AWG (A) is connected with a single-mode fiber (B) at the input and fixed on an isothermal plate (C). The output waveguides end facets are attached directly to the vertical CMOS linear image sensor (D). The CMOS sensor is read-out by a microcontroller (E). The TEC controller (F) is also allocated within the system.

2.3. Fiber Bragg Grating Fabrication

For the production of the FBG, a femtosecond pulsed laser-based point-by-point inscription method is applied. With a three-dimensional computer-controlled translation stage (N-565.260 linear translation stage, Physik Instrumente (PI), Karlsruhe, Germany) and an objective lens (LD Plan-Neoflur 20×, Zeiss, Oberkochen, Germany), the utilized standard telecommunication single-mode glass fiber (SMF810-E5/125PI, Leoni, Nuremberg, Germany) is optically focused on its light-guiding core. During processing, pulses from a femtosecond laser (Ti:Sapphire Tsunami/Spitfire pro, Spectra-Physics, Santa Clara, CA, USA) pass the lens and form single grating points with a locally increased refractive index. The fiber is moved by the translation stage after each pulse until the entire type II fiber Bragg grating is produced. Due to the adaptability of this manufacturing method, the properties of the gratings can be systematically customized to the requirements of the respective polymer AWG channels. For the current experimental version, a Bragg grating with a central reflection wavelength of 852.30 nm, a peak width at half-height of 0.90 nm and a reflectivity of 90.0% is manufactured (figure of spectrum in the appendix). To suppress interfering secondary modes, linear shape apodization is used. With a grating length of 1.15 mm comprising of 700 single refractive index modifications, a reflection spectrum approximately shaped like an ideal Gaussian curve can thus be generated. This shape is particularly suitable for calibrating the AWG channels as described in the following section.

2.4. System Calibration

In order to obtain the correlation between light intensity at the end of the output waveguides and the central wavelength of the narrowband spectrum reflected by the FBG at the AWG input, a wavelength calibration is done, and the results are shown in Figure 3. The setup for the system calibration is largely similar to the one for the battery experiments that are presented in the following section. It mainly consists of the FBG, the AWG, an optical spectrum analyzer (OSA) (AQ6373B, Yokogawa, Tokyo, Japan) and a superluminescent diode (SLED) (EXS210037-01, Exalos, Schlieren, Switzerland). The major difference is that the FBG is not fixed to the battery cell at this point. Instead, the FBG, described in the previous section, is fixed with one end to a manual translational stage (NanoMax-TS Max302/M, Thorlabs, Newton, MA, USA) and with the other end to a rigid post. At a room temperature of 21 °C, the FBG is randomly stretched by incrementally moving the translational stage, resulting in a shift of the reflected wavelength between 852.5 nm and 853.25 nm. The broadband light spectrum of the SLED is launched into the fiber beginning and partly reflected at the FBG from where it is guided via a 3dB coupler (FC850-40-50-APC, Thorlabs, Newton, MA, USA) to the input of the AWG. The remaining spectrum, reduced by the reflected spectral range, is transmitted to the end of the fiber, where the actual FBG wavelength is evaluated by the OSA. This reference value is stored together with the intensities at the three AWG output waveguides, that as a result of the reflected FBG signal at the input, are captured by the CMOS sensor. To minimize possible errors, due to AWG waveguide outputs positioned between two pixels of the CMOS linear image sensor, the intensities of the pixel before and after the aforementioned pixels (407, 486, 565) are also taken into account and an averaged value is calculated.

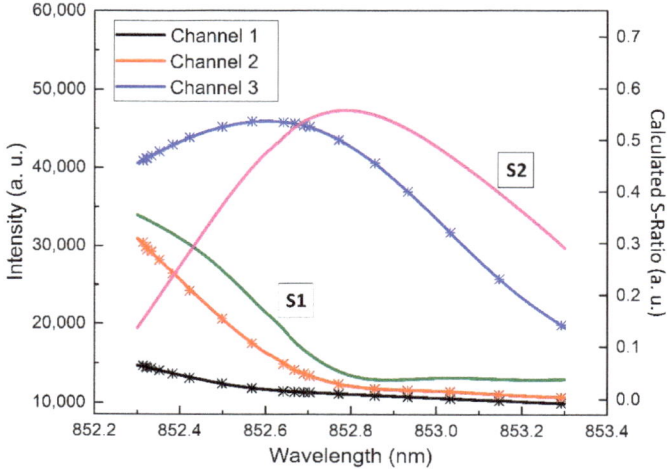

Figure 3. Correlation between center wavelength of the light portion reflected at the fiber Bragg grating (FBG) and the averaged intensities of the CMOS pixels at the AWG output waveguide positions. By using the fitted values as inputs for Equation (1), the green S-ratio course was calculated for channel 1 and 2 and the magenta S-ratio course was calculated for channel 2 and 3, respectively.

Subsequently, the intensities are used to interpolate the course of the AWG output signals with respect to the center wavelength of the light portion reflected by FBG. Finally, according to [37], the ratio between the difference of two adjacent channel intensities over their sum is calculated, as shown in Equation (1), where I_i is the averaged intensity of channel i.

$$S_i = (I_{i+1} - I_i)/(I_{i+1} + I_i) \tag{1}$$

The ratio S is calculated depending on the reflected wavelength and stored with a resolution of 1 pm as a look-up table in the analysis software, in order to obtain an expression for the wavelength that is independent from the power of the light source as well as from the integration time of the CMOS image sensor. The integration time for all experiments presented here is set to 10 ms and usually 50 scans were averaged, thus an overall data acquisition frequency of 2 Hz results.

3. Results and Discussion

To demonstrate the performance of the new polymeric AWG interrogator, long-term experiments with the presented FBG sensor were conducted. The optical sensor, in this particular case, serves predominately as a strain sensor since it is fixed to the surface of a lithium-ion pouch cell by means of instant adhesive. Although, FBG sensors are sensitive to both temperature and strain, the reader should note that for the cyclization experiments presented in this paper, the shift in reflected wavelength is almost exclusively due to strain because the current rate is low and the resulting heat generation of the cell minimal. Furthermore, the cell is placed in a temperature regulated environment, first to investigate the shift of the reflected wavelength due to changes of the temperature (part A) and second to ensure a constant temperature during the cycling experiments (part B). The AWG interrogator is placed along with the light source and the coupler in a separate temperature chamber that is set to a constant temperature of 16 °C. The intensity signals transmitted from the microcontroller are processed by a personal computer in the analysis software in which the previously obtained relationship between intensities and FBG wavelength is calculated, displayed and finally stored. The overall experimental setup is schematically shown in Figure 4.

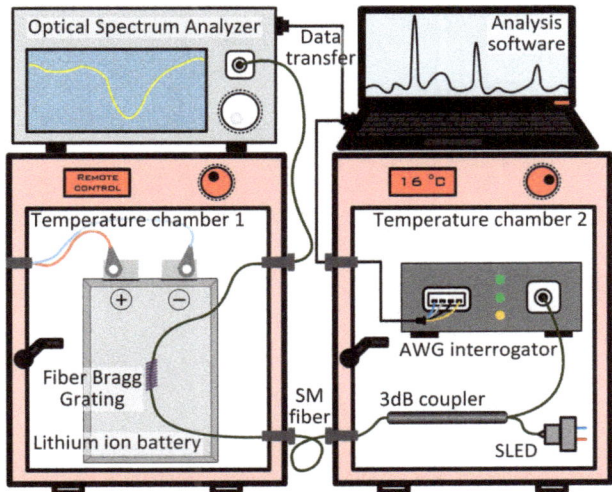

Figure 4. General setup for the temperature (part A) and cyclic experiments (part B). The AWG interrogator, along with the superluminescent diode (SLED) light source, and the fiber-coupler are put into a temperature chamber with a constant temperature of 16 °C and a constant relative humidity of 43% (only for part B). The investigated lithium-ion cell is put in a temperature chamber with a variable temperature for experimental part A and a constant temperature of 20 °C for experimental part B, respectively. The cell is electrically connected to a battery cyclization unit and optically connected to the AWG as well as to the optical spectrum analyzer (OSA). The AWG evaluates the FBG wavelength through the reflected spectrum and the OSA evaluates it through the transmitted spectrum.

3.1. Temperature and Humidity Variation

In order to ascertain a proper fixing of the FBG to the lithium-ion cell surface and to gain information about the influence of the cell temperature as well as of the relative air humidity on the reflected wavelength, first, the cell is exposed to temperature variations. For this purpose, the temperature chamber 1 is set to 20 °C for a sufficiently long time, followed by a step to 25 °C, and after a rest time of 4 h, the temperature is increased by a step of 5 K again. The temperature is held at 30 °C for 4 h and afterwards the temperature is decreased to 20 °C again, by means of 5 K steps, with the same resting periods as during temperature raise. These steps are conducted two times, followed by 40 h with constant temperature. The results can be seen in Figure 5.

During the experiment, relative humidity of the surrounding air in the temperature chamber 2 is additionally recorded with a digital sensor (HYT 939, Innovative Sensor Technology, Ebnat-Kappel, Switzerland). As a result of the regional weather changes, the values for the relative air humidity vary between 45%–55% inside the non-air-conditioned laboratory. For the isothermal period starting from experiment hour 30, no change of the reflected wavelength is expected and almost none is measured by the OSA. Nevertheless, the variation of the reflected wavelength measured by the AWG is 110 pm, as shown in Figure 5.

It is well known that polymeric plastics in general and the EpoClad/EpoCore photoresists used within this research work in particular are hygroscopic [38], and the absorption of water molecules, in turn, changes the optical properties of the AWG. In Figure 6, it can be seen, that for the herein presented humidity range, a linear relationship, including a hysteresis, exists.

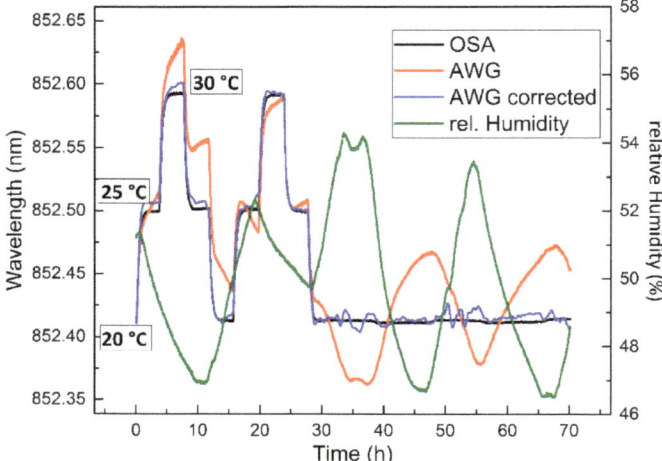

Figure 5. FBG wavelength measured by the OSA and the AWG, respectively, during the temperature experiment together with the relative air humidity in the surrounding of the AWG. The humidity strongly influences the optical properties of the AWG. The corrected AWG values (blue course) are calculated by using Equation (2).

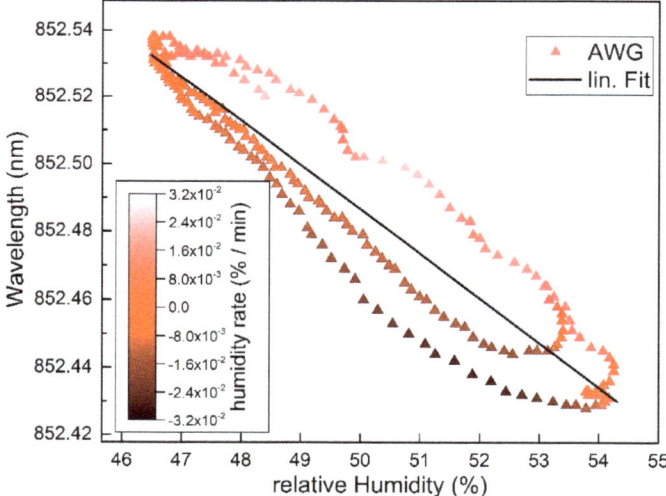

Figure 6. FBG wavelength measured by the AWG during the isothermal period of the temperature experiment shown in Figure 5. In the relevant humidity range there is a linear correlation between the optical output of the AWG and the relative humidity.

To minimize the variation of the AWG output due to the change of relative humidity, a multiple linear regression fit is done, expressed by Equation (2), where ξ is the relative air humidity in percent and $\dot{\xi}$ the change of the relative humidity in %/min. Λ is the regression constant determined to 853.1 nm, α and β are coefficients determined to -0.0128 nm/% and 0.9077 nm·min/%, respectively. Λ_c, as the result of Equation (2), represents the corrected wavelength, shown as the blue course in Figure 5.

$$\Lambda_c(\xi, \dot{\xi}) = \Lambda + \alpha\,\xi + \beta\,\dot{\xi} \qquad (2)$$

The disturbance-related variation of the AWG values can be decreased to 20 pm during the isothermal period by applying Equation (2). The deviation from the values measured by the OSA during change of the temperature is also improved.

3.2. State of Charge Variation

Although the influence of the relative air humidity is known and can be minimized, the relative air humidity in temperature chamber 2 is kept constant (43.4% ± 0.9%) by using a vessel with a saturated salt solution of potassium carbonate (K_2CaO_3) for the long-term cyclization experiment. Furthermore, the temperature of the temperature chamber 1 is set to 20 °C during the whole time, thus the Bragg sensor is particularly sensitive to changes of the cell's surface strain.

In Figure 7, the result of the cyclic experiment is shown. For a time of 27 days, the lithium-ion cell underwent 25 full charge–discharge cycles with a current of 5 A between 4.2 V and 3.0 V. It can be seen that the reflected wavelength signal, measured by the AWG, is in good agreement to the signal measured by the OSA in the transmitted spectrum.

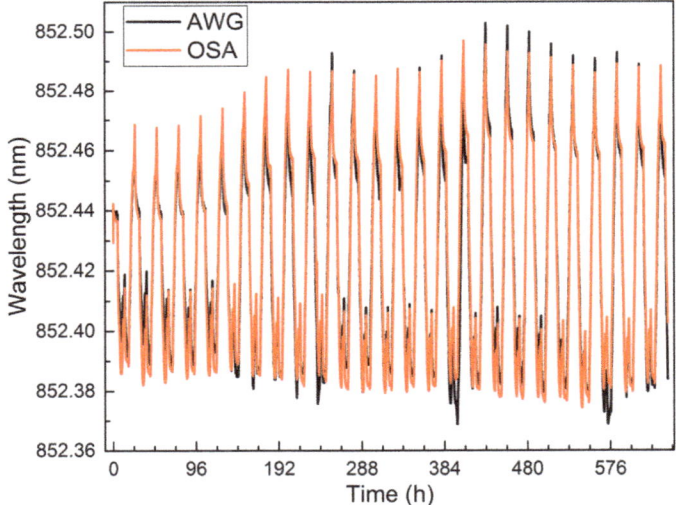

Figure 7. Result of the cyclization experiment. Over 27 days, 25 cycles are performed. The AWG signal is in good agreement to the OSA signal at any time and represents the battery status well.

For every time step, the error of the AWG is shown in Figure 8, along with the values for the relative humidity. Even though the maximum error is in the range of 3×10^{-2} nm during the 16th cycle (at experiment time 390), the total mean error is 6.5×10^{-4} nm, with a standard deviation of 5.9×10^{-3} nm. Furthermore, it is evident that the largest deviation between OSA and AWG occur when there are significant variations of the relative humidity, like in hour 240, 408 and 576, respectively.

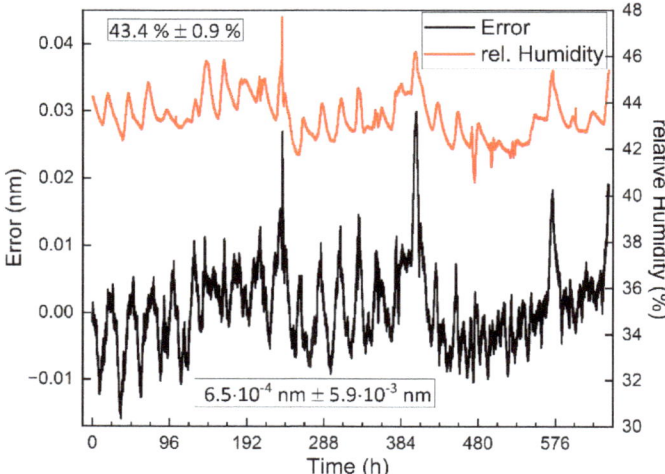

Figure 8. Difference between AWG and OSA signal together with the relative humidity. The largest errors occur when there are great variations in the humidity.

From this long-term experiment, an estimation of both the accuracy and precision of our system can be derived. At the selected settings, our Yokogawa AQ6373B optical spectrum analyzer features a wavelength accuracy of $\pm 5 \times 10^{-2}$ nm and a resolution of 2×10^{-2} nm. The above-described error between OSA and AWG adds to that, so that the accuracy of our system can be calculated to $\pm 7.59 \times 10^{-2}$ nm.

To demonstrate the eligibility of the presented polymeric AWG interrogator as a useful device for the status monitoring of lithium-ion cells, one single cell charge–discharge cycle is shown in detail in Figure 9. The cell charging starts at experiment hour 601 when the cell voltage rises significantly and the first drop in the wavelength takes place. This is caused by a typical temperature decrease at the beginning of the charge process, since endothermic chemical processes are predominant over the Joule heat generation. In the ongoing course, the reflected wavelength begins to rise, as the lithium-ions intercalate to the anode, causing an increase of the cell's volume and therefore of the surface strain. Reference [39] demonstrates the correlation between the graphite anode potential (vs. Li/Li$^+$) of a lithium-ion battery and its intercalation stages. This can be seen in Figure 9 by different voltage rates of change that are typical for lithium-ion cells. The optical signal also has a nonlinear course and shows different rates of change, which makes them suitable for the detection of characteristic phase transition points that can be used for status monitoring of lithium-ion batteries, although the exact correlation between voltage signal and volumetric behavior is not fully understood yet.

The charging terminates at the signal peak at experiment hour 609.5, followed by a rest period of 5 h, during which a relaxation of the cell takes place. The voltage signal decreases only slightly but a larger decrease in the strain signal occurs, showing ion diffusion processes on the one side and a temperature approximation to the ambient temperature on the other side. It is known that this behavior can also change with an ongoing degradation of the cell because the open-circuit voltage of a lithium-ion cell is linked to its capacity, what can be used to determine the actual SOC or SOH, for example by performing incremental capacity analysis [40].

The discharge cycle starts at hour 615 and is identifiable by a sudden decrease of the voltage signal and the reflected wavelength. Similar to the charging period, the course of the voltage signal shows again characteristic rate changes at certain points caused by the deintercalation of the lithium-ions from the graphite anode and it can be seen in Figure 9 that the strain course also changes at these points. At the end of the discharge cycle at experiment hour 619, the signal rises again, which is due to a significant temperature rise, caused by a typically increasing internal cell resistance that leads to increased Joule heating. After discharge, the cell rests again and the voltage converges to its open circuit voltage. The strain signal decreases with decreasing cell temperature and converges to its initial value.

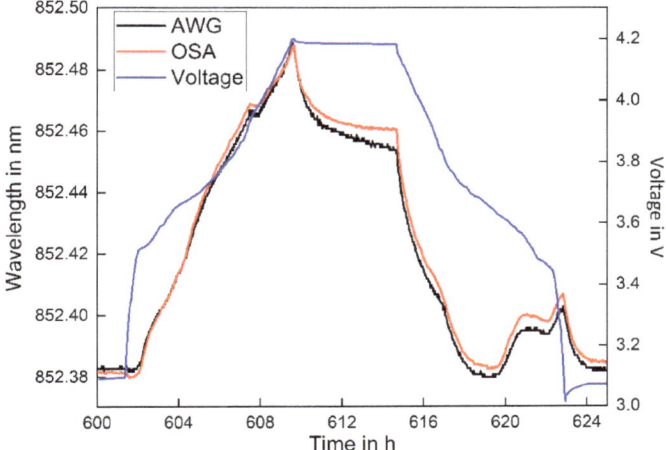

Figure 9. Exemplary single cycle showing the typical optically measured strain behavior of a lithium-ion cell along with the cell voltage.

4. Conclusions

With this work, we present a new, precise and cost-effective approach to read-out FBG utilizing a novel AWG interrogation system. The polymeric AWG are simple and cost-efficient to produce by using direct laser lithography, a technique that allows rapid development of integrated optical systems and enables short times from simulation to an operational prototype. For the application as a status monitoring device of lithium-ion batteries, the designed AWG interrogation system shows good optical performance. With a comparison to an optical spectrum analyzer, we calculated the approximate accuracy of our new measurement system to $\pm 7.59 \times 10^{-2}$ nm. The resolution is defined by the S-functions to 1×10^{-3} nm. The obtained accuracy is sufficient to observe the typical strain behavior (e.g., as in ref. [25,26]) of a single cell during one full charge cycle and was maintained for 25 cycles over one month. Nevertheless, the influence of the relative air humidity is not negligible and has to be investigated further. Although we are able to minimize the error with simple methods, the systems robustness has to be improved in order to become competitive to state-of-the-art electrical BMS. Besides this, future work will focus on AWG with at least 16 output channels to distinguish up to 8 FBG and to enable the usage of a reference FBG for temperature compensation in order to monitor a battery cell under non standardized conditions and extend the observations to a multitude of cells. The herein presented interrogation unit can furthermore be used to evaluate a multitude of FBG by using standard optical accessories e.g., an optical switch, or by using a multilayer design with several stacked AWG in combination with a two-dimensional CMOS image sensor to evaluate the individual output channel intensities. In future studies, the acquired optical information will be investigated in more detail for use in advanced algorithms that are able to exactly determine the SOC and SOH of a battery system. Furthermore, it is desirable to develop an all polymeric sensor system that consists not only of a polymeric AWG, but also takes advantage of other polymeric components, like a polymeric coupler and FBG.

Author Contributions: For the presented research work the authors are contributed in the following manner: J.M., A.N. and W.S. are responsible for the conceptualization and funding acquisition; J.M. and A.N. are responsible for data curation, investigation, methodology, visualization and writing; J.M. is responsible for formal analysis and software; A.N. and C.K. are responsible for project administration; E.P., C.K., A.N. and J.M. are responsible for resources; C.K. is responsible for validation; C.K., A.N. and J.M. are responsible for writing—review and editing; W.S is responsible for supervision.

Funding: This research was funded by the German Federal Ministry for Economic Affairs and Energy, grant number 03ET6105A.

Acknowledgments: Financial support for the conduction of the experiments from the Federal Ministry for Economic Affairs and Energy is gratefully acknowledged.

Conflicts of Interest: The authors declare no conflict of interest.

References

1. Li, M.; Lu, J.; Chen, Z.; Amine, K. 30 Years of Lithium-Ion Batteries. *Adv. Mater. Weinh.* **2018**, *30*, e1800561. [CrossRef] [PubMed]
2. Zubi, G.; Dufo-López, R.; Carvalho, M.; Pasaoglu, G. The lithium-ion battery: State of the art and future perspectives. *Renew. Sustain. Energy Rev.* **2018**, *89*, 292–308. [CrossRef]
3. Cano, Z.P.; Banham, D.; Ye, S.; Hintennach, A.; Lu, J.; Fowler, M.; Chen, Z. Batteries and fuel cells for emerging electric vehicle markets. *Nat. Energy* **2018**, *3*, 279–289. [CrossRef]
4. Telaretti, E.; Dusonchet, L. Stationary battery systems in the main world markets: Part 1: Overview of the state-of-the-art. In Proceedings of the 2017 IEEE International Conference on Environment and Electrical Engineering and 2017 IEEE Industrial and Commercial Power Systems Europe (EEEIC/I&CPS Europe), Milan, Italy, 6–9 June 2017; pp. 1–7, ISBN 978-1-5386-3917-7.
5. De Sisternes, F.J.; Jenkins, J.D.; Botterud, A. The value of energy storage in decarbonizing the electricity sector. *Appl. Energy* **2016**, *175*, 368–379. [CrossRef]
6. Abada, S.; Marlair, G.; Lecocq, A.; Petit, M.; Sauvant-Moynot, V.; Huet, F. Safety focused modeling of lithium-ion batteries: A review. *J. Power Sources* **2016**, *306*, 178–192. [CrossRef]
7. Feng, X.; Ouyang, M.; Liu, X.; Lu, L.; Xia, Y.; He, X. Thermal runaway mechanism of lithium ion battery for electric vehicles: A review. *Energy Storage Mater.* **2018**, *10*, 246–267. [CrossRef]
8. Federal Aviation Administration. Aviation Cargo and Passenger Baggage Events Involving Smoke, Fire, Extreme Heat or Explosion Involving Lithium Batteries or Unknown Battery Types. 2018. Available online: https://www.faa.gov/hazmat/resources/lithium_batteries/media/Battery_incident_chart.pdf (accessed on 26 March 2019).
9. Monhof, M.; Beverungen, D.; Klör, B.; Bräuer, S. Extending Battery Management Systems for Making Informed Decisions on Battery Reuse. In *New Horizons in Design Science: Broadening the Research Agenda*; Donnellan, B., Helfert, M., Kenneally, J., VanderMeer, D., Rothenberger, M., Winter, R., Eds.; Springer International Publishing: Cham, Switzerland, 2015; pp. 447–454. ISBN 978-3-319-18713-6.
10. Rohr, S.; Wagner, S.; Baumann, M.; Muller, S.; Lienkamp, M. A techno-economic analysis of end of life value chains for lithium-ion batteries from electric vehicles. In Proceedings of the 2017 Twelfth International Conference on Ecological Vehicles and Renewable Energies (EVER), Monte-Carlo, Monaco, 11–13 April 2017; pp. 1–14, ISBN 978-1-5386-1692-5.
11. Andrea, D. *Battery Management Systems for Large Lithium-Ion Battery Packs*; Artech House: Boston, MA, USA, 2010; ISBN 978-1-60807-104-3.
12. Hannan, M.; Lipu, M.; Hussain, A.; Mohamed, A. A review of lithium-ion battery state of charge estimation and management system in electric vehicle applications: Challenges and recommendations. *Renew. Sustain. Energy Rev.* **2017**, *78*, 834–854. [CrossRef]
13. Xiong, R.; Li, L.; Tian, J. Towards a smarter battery management system: A critical review on battery state of health monitoring methods. *J. Power Sources* **2018**, *405*, 18–29. [CrossRef]
14. Cannarella, J.; Arnold, C.B. Stress evolution and capacity fade in constrained lithium-ion pouch cells. *J. Power Sources* **2014**, *245*, 745–751. [CrossRef]
15. Lee, B. Review of the present status of optical fiber sensors. *Opt. Fiber Technol.* **2003**, *9*, 57–79. [CrossRef]
16. Chen, J.; Liu, B.; Zhang, H. Review of fiber Bragg grating sensor technology. *Front. Optoelectron. China* **2011**, *4*, 204–212. [CrossRef]

17. Chikh-Bled, H.; Chah, K.; Álvaro, G.-V.; Caucheteur, C.; Lasri, B. Behavior of femtosecond laser-induced eccentric fiber Bragg gratings at very high temperatures. *Opt. Lett.* **2016**, *41*, 4048. [CrossRef]
18. Meyer, J.; Nedjalkov, A.; Doering, A.; Angelmahr, M.; Schade, W. Fiber optical sensors for enhanced battery safety. In Proceedings of the Fiber Optic Sensors and Applications XII. SPIE Sensing Technology + Applications, Baltimore, MD, USA, 20–24 April 2015. [CrossRef]
19. Nascimento, M.; Ferreira, M.S.; Pinto, J.L. Real time thermal monitoring of lithium batteries with fiber sensors and thermocouples: A comparative study. *Measurement* **2017**, *111*, 260–263. [CrossRef]
20. Yang, G.; Leitão, C.; Li, Y.; Pinto, J.; Jiang, X. Real-time temperature measurement with fiber Bragg sensors in lithium batteries for safety usage. *Measurement* **2013**, *46*, 3166–3172. [CrossRef]
21. Novais, S.; Nascimento, M.; Grande, L.; Domingues, M.F.; Antunes, P.; Alberto, N.; Leitão, C.; Oliveira, R.; Koch, S.; Kim, G.T.; et al. Internal and External Temperature Monitoring of a Li-Ion Battery with Fiber Bragg Grating Sensors. *Sensors* **2016**, *16*, E1394. [CrossRef]
22. Raghavan, A.; Kiesel, P.; Sommer, L.W.; Schwartz, J.; Lochbaum, A.; Hegyi, A.; Schuh, A.; Arakaki, K.; Saha, B.; Ganguli, A.; et al. Embedded fiber-optic sensing for accurate internal monitoring of cell state in advanced battery management systems part 1: Cell embedding method and performance. *J. Power Sources* **2017**, *341*, 466–473. [CrossRef]
23. Lochbaum, A.; Kiesel, P.; Sommer, L.W.; Bae, C.-J.; Staudt, T.; Saha, B.; Raghavan, A.; Lieberman, R.; Delgado, J.; Choi, B.; et al. Embedded Fiber Optic Chemical Sensing for Internal Cell Side-Reaction Monitoring in Advanced Battery Management Systems. *MRS Proc.* **2014**, *1681*, 715. [CrossRef]
24. Nedjalkov, A.; Meyer, J.; Gräfenstein, A.; Schramm, B.; Angelmahr, M.; Schwenzel, J.; Schade, W. Refractive Index Measurement of Lithium Ion Battery Electrolyte with Etched Surface Cladding Waveguide Bragg Gratings and Cell Electrode State Monitoring by Optical Strain Sensors. *Batteries* **2019**, *5*, 30. [CrossRef]
25. Bae, C.-J.; Manandhar, A.; Kiesel, P.; Raghavan, A. Monitoring the Strain Evolution of Lithium-Ion Battery Electrodes using an Optical Fiber Bragg Grating Sensor. *Energy Technol.* **2016**, *4*, 851–855. [CrossRef]
26. Sommer, L.W.; Kiesel, P.; Ganguli, A.; Lochbaum, A.; Saha, B.; Schwartz, J.; Bae, C.-J.; Alamgir, M.; Raghavan, A. Fast and slow ion diffusion processes in lithium ion pouch cells during cycling observed with fiber optic strain sensors. *J. Power Sources* **2015**, *296*, 46–52. [CrossRef]
27. Sommer, L.W.; Raghavan, A.; Kiesel, P.; Saha, B.; Schwartz, J.; Lochbaum, A.; Ganguli, A.; Bae, C.-J.; Alamgir, M. Monitoring of Intercalation Stages in Lithium-Ion Cells over Charge-Discharge Cycles with Fiber Optic Sensors. *J. Electrochem. Soc.* **2015**, *162*, A2664–A2669. [CrossRef]
28. Tosi, D. Review and Analysis of Peak Tracking Techniques for Fiber Bragg Grating Sensors. *Sensors* **2017**, *17*, 2368. [CrossRef]
29. Jackson, D.A.; Bennion, I.; Rao, Y.J.; Zhang, L. Dual-cavity interferometric wavelength-shift detection for in-fiber Bragg grating sensors. *Opt. Lett.* **1996**, *21*, 1556. [CrossRef]
30. Tsao, S.-L.; Peng, P.-C. An SOI Michelson interferometer sensor with waveguide Bragg reflective gratings for temperature monitoring. *Microw. Opt. Technol. Lett.* **2001**, *30*, 321–322. [CrossRef]
31. Kersey, A.D.; Berkoff, T.A.; Morey, W.W. Multiplexed fiber Bragg grating strain-sensor system with a fiber Fabry-Perot wavelength filter. *Opt. Lett.* **1993**, *18*, 1370. [CrossRef]
32. Jung, E.J.; Kim, C.-S.; Jeong, M.Y.; Kim, M.K.; Jeon, M.Y.; Jung, W.; Chen, Z. Characterization of FBG sensor interrogation based on a FDML wavelength swept laser. *Opt. Express* **2008**, *16*, 16552–16560.
33. Talahashi, H.; Oda, K.; Toba, H.; Inoue, Y.; Takahashi, H. Transmission characteristics of arrayed waveguide N×N wavelength multiplexer. *J. Light. Technol.* **1995**, *13*, 447–455. [CrossRef]
34. Sano, Y.; Yoshino, T. Fast optical wavelength interrogator employing arrayed waveguide grating for distributed fiber bragg grating sensors. *J. Light. Technol.* **2003**, *21*, 132–139. [CrossRef]
35. Tsuchizawa, T.; Yamada, K.; Watanabe, T.; Park, S.; Nishi, H.; Kou, R.; Shinojima, H.; Itabashi, S.-I. Monolithic Integration of Silicon-, Germanium-, and Silica-Based Optical Devices for Telecommunications Applications. *IEEE J. Select. Top. Quantum Electron.* **2011**, *17*, 516–525. [CrossRef]
36. Pichler, E.; Bethmann, K.; Kelb, C.; Schade, W. Rapid prototyping of all-polymer AWGs for FBG readout using direct laser lithography. *Opt. Lett.* **2018**, *43*, 5347–5350. [CrossRef]
37. Koch, J.; Angelmahr, M.; Schade, W. Arrayed waveguide grating interrogator for fiber Bragg grating sensors: Measurement and simulation. *Appl. Opt.* **2012**, *51*, 7718–7723. [CrossRef]
38. Gijsenbergh, P.; Wouters, K.; Vanstreels, K.; Puers, R. Determining the physical properties of EpoClad negative photoresist for use in MEMS applications. *J. Micromech. Microeng.* **2011**, *21*, 74001. [CrossRef]

39. Sethuraman, V.A.; Hardwick, L.J.; Srinivasan, V.; Kostecki, R. Surface structural disordering in graphite upon lithium intercalation/deintercalation. *J. Power Sources* **2010**, *195*, 3655–3660. [CrossRef]
40. Weng, C.; Sun, J.; Peng, H. A unified open-circuit-voltage model of lithium-ion batteries for state-of-charge estimation and state-of-health monitoring. *J. Power Sources* **2014**, *258*, 228–237. [CrossRef]

© 2019 by the authors. Licensee MDPI, Basel, Switzerland. This article is an open access article distributed under the terms and conditions of the Creative Commons Attribution (CC BY) license (http://creativecommons.org/licenses/by/4.0/).

Article

Sensor Fault Detection and Isolation for Degrading Lithium-Ion Batteries in Electric Vehicles Using Parameter Estimation with Recursive Least Squares

Manh-Kien Tran and Michael Fowler *

Department of Chemical Engineering, University of Waterloo, Waterloo, ON N2L 3G1, Canada; kmtran@uwaterloo.ca
* Correspondence: mfowler@uwaterloo.ca; Tel.: +1-(519)-888-4567 (ext. 33415)

Received: 4 October 2019; Accepted: 17 December 2019; Published: 20 December 2019

Abstract: With the increase in usage of electric vehicles (EVs), the demand for Lithium-ion (Li-ion) batteries is also on the rise. The battery management system (BMS) plays an important role in ensuring the safe and reliable operation of the battery in EVs. Sensor faults in the BMS can have significant negative effects on the system, hence it is important to diagnose these faults in real-time. Existing sensor fault detection and isolation (FDI) methods have not considered battery degradation. Degradation can affect the long-term performance of the battery and cause false fault detection. This paper presents a model-based sensor FDI scheme for a Li-ion cell undergoing degradation. The proposed scheme uses the recursive least squares (RLS) method to estimate the equivalent circuit model (ECM) parameters in real time. The estimated ECM parameters are put through weighted moving average (WMA) filters, and then cumulative sum control charts (CUSUM) are implemented to detect any significant deviation between unfiltered and filtered data, which would indicate a fault. The current and voltage faults are isolated based on the responsiveness of the parameters when each fault occurs. The proposed FDI scheme is then validated through conducting a series of experiments and simulations.

Keywords: fault detection and isolation; sensor fault; battery model; battery management systems; battery degradation; electric vehicles; online parameter estimation; recursive least squares

---

1. Introduction

Lithium-ion (Li-ion) batteries are the most popular form of energy storage in the world, amounting to 85.6% of energy storage systems utilized in 2015. Although it has the highest price, it shows the lowest cost per cycle [1]. The substantial demand for Li-ion batteries is due to portable devices and electric vehicles (EVs). Li-ion batteries are used in EVs due to their high power and energy density, long life span, and low environmental impact. EVs require a battery system that consists of hundreds or thousands of single cells. In order to manage this large number of cells, the battery pack needs a battery management system (BMS). It is important that the performance of the BMS is accurate and reliable, to ensure the performance and safety in EVs application. The functions of the BMS include state of charge (SOC) and state of health (SOH) estimation, and over-current and over-voltage protection [2]. These functions rely heavily on voltage and current sensor measurements [3]. It is possible for the sensors to experience malfunctions during the operation of the battery, due to manufacturing defects or environmental factors. The estimation of the SOC (similar to a fuel meter in conventional vehicles) and the SOH (similar to an odometer), would be affected if there were any faults with the sensors, leading to over-charge and/or over-discharge phenomenon which would degrade the battery faster. The current and voltage protection would also fail to work properly due to faulty sensors. This can lead to more catastrophic failures since the current and voltage can exceed their operational limits undetected, due to incorrect sensor readings [4]. Even though a sensor fault with a small magnitude does not immediately affect the battery performance, it can have a significant effect over time. This can be prevented by detecting and resolving the sensor fault promptly after it develops. Although the authors are not aware of any published data on the failure rates of BMS sensors in EVs, it is reasonable to anticipate some failures due to the nature of the application. The sensors are subject to vibration and physical damage from collisions, which can ultimately lead to disconnection or resistance build-up of the wires and cause deviations in the readings. Therefore, it is critical to develop an algorithm that can reliably and accurately diagnose any faulty operation of the voltage and current sensors in real time.

The reviews on fault mechanism and diagnosis approaches for Li-ion batteries can be found in [2,5]. Desirable characteristics of a fault detection and isolation (FDI) scheme include quick detection and diagnosis, isolability, robustness, adaptability, low modelling requirements, and a reasonable balance between storage and computational requirements [6]. Several existing FDI methods were able to accomplish some of the desired characteristics stated above. An extended Kalman filter was used in [4] to diagnose sensor faults, but fault isolation was not achieved. This study confirms that the battery can be over-charged or over-discharged due to sensor faults, caused by the inaccuracy of SOC estimation. In [7], the nonlinear parity equation approach, coupled with sliding mode observers, were used to develop an FDI scheme to detect sensor faults for a single battery cell. A set of Luenberger and learning observers were used in [8] for simultaneous single-fault isolation and estimation of a faulty cell in a battery string. In [9], an FDI strategy using structural analysis theory and statistical inference residual evaluation was presented, but the computational effort was rather high. An FDI scheme using sliding mode observers with equivalent output error injection was introduced in [10], with findings that show false detection rate is affected by the variation in model parameters. All of the methods mentioned above work under the assumption that the battery model parameters remain constant throughout the battery pack's life span. However, the parameters can be affected by degradation, a significant property of battery operation. There has not been any mention of cell degradation in any FDI works or literature.

There are a few models used to illustrate battery behavior, but the equivalent circuit model (ECM) is the most widely used in FDI works [5]. The parameters of the ECM were derived using conservation of species, conservation of charge, and reaction kinetics in [11]. The results show that the parameters have physical meanings and can be affected by the chemistry of the battery, as well as the environment of operation. Therefore, degradation of the battery would have some effects on the parameters. The existing FDI schemes can be improved by integrating degradation into the ECM. However, this has been proven to be a difficult task. Currently, battery degradation models can be obtained by fitting experimental data under constant conditions. However, this is not an appropriate model for battery degradation in EVs applications, due to its complex operating state [12]. Experimental models are also less accurate, time-consuming, and costly. Adaptive models are more accurate, but require training to estimate the parameters that correlate with degradation. Moreover, the models can have high computational effort which is not suitable for real-time BMS applications [13]. Another approach is needed to effectively diagnose faults while considering the effect of degradation on ECM parameters, which this paper will present.

The key contribution of this paper is the proposal of a model-based sensor FDI scheme for Li-ion battery in EVs while considering battery degradation. The ECM parameters are expected to change during battery operation due to the effect of degradation. The paper studies and confirms this effect through a series of experiments. The proposed FDI scheme uses the recursive least squares (RLS) method to estimate the ECM parameters in real time, then applies a weighted moving average (WMA) filter coupled with a cumulative sum control chart (CUSUM) to detect any voltage and current sensor faults. The use of RLS is suggested because of its low computational demand and easy implementation [14]. The implementation of the WMA filter eliminates the concern of battery degradation, in addition to the effect of SOC and temperature on ECM parameters. Furthermore, the sensor faults are isolated based on the responsiveness of the parameters when a specific fault occurs. Finally, the Urban Dynamometer Driving Schedule (UDDS) cycle with sensor fault simulation is applied to validate and evaluate the performance of the proposed FDI scheme for a lithium iron phosphate (LFP) cell.

The rest of this paper is organized as follows: Section 2 describes the battery model used for this work, while Section 3 outlines the details of the proposed FDI scheme. Section 4 provides the experimental design and analysis of the effect of degradation and various faults on the parameters. The evaluation of the proposed fault diagnosis scheme is presented in Section 5, and the resulting conclusions are given in Section 6.

2. Battery Modelling

The most common model used to describe battery behaviors in EVs application is the equivalent circuit model. For an LFP battery running drive cycles that are highly dynamic, such as UDDS, an ECM with at least two RC pairs is recommended [15]. This is because the first order ECM neglects the effect of diffusion. However, the higher the model order is, the more computational effort it demands, due to the larger number of model parameters. For the implementation of the proposed FDI, it is not required for the model to have great accuracy, since the extraction of ECM parameters is used to monitor the state of battery operation, rather than to model the battery performance. Therefore, in order to optimize the computational complexity of the approach, the first order ECM is used in this paper. The simplified ECM model is shown in Figure 1.

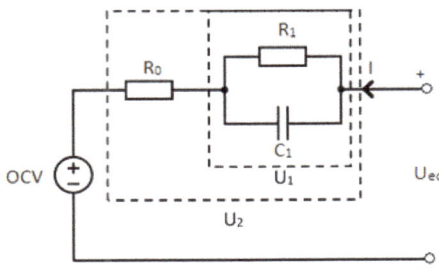

Figure 1. Schematic of a first order equivalent circuit model (ECM).

The state space equation of this battery model can be expressed as follows:

$$\begin{aligned} \dot{U}_1 &= \frac{I}{C_1} - \frac{U_1}{C_1 R_1} \\ U_{eq} &= OCV - U_1 - IR_0 \end{aligned} \quad (1)$$

In order to perform the proposed recursive approach on the model, an autoregressive exogenous model is needed. This is done through obtaining the transfer function of the battery impedance from Equation (1) in the s-domain, as shown in Equation (2). The transfer function is then discretized using the basic forward Euler transformation method, in which s is replaced by $\frac{1-z^{-1}}{T \cdot z^{-1}}$, where T is the sampling time. The discretization is shown in Equation (3) below.

$$G(s) = \frac{U_2(s)}{I(s)} = -R_0 - \frac{R_1}{1 + sR_1 C_1} \quad (2)$$

$$G(z) = \frac{a_2 + a_3 z^{-1}}{1 + a_1 z^{-1}} \quad (3)$$

where

$$a_1 = \frac{T}{R_1 C_1} - 1 \quad (4)$$

$$a_2 = -R_0 \tag{5}$$

$$a_3 = R_0 - \frac{T}{C_1} - \frac{TR_0}{R_1 C_1} \tag{6}$$

R_1 and C_1 can be determined as follows:

$$R_1 = \frac{a_1 a_2 - a_3}{1 + a_1} \tag{7}$$

$$C_1 = \frac{T}{a_1 a_2 - a_3} \tag{8}$$

The autoregressive exogenous model can then be obtained as follows:

$$y_k = OCV_k + a_1(OCV_{k-1} - y_{k-1}) + I_k a_2 + I_{k-1} a_3 \tag{9}$$

with y_k, which can be rewritten as:

$$y_k = \theta_k^T \phi_k \tag{10}$$

where

$$\theta_k = [1; a_{1,k}; a_{2,k}; a_{3,k}] \tag{11}$$

$$\phi_k = [OCV_k; (OCV_{k-1} - y_{k-1}); I_k; I_{k-1}] \tag{12}$$

The values for OCV (open-circuit voltage) in Equation (12) will be determined from the OCV–SOC relationship, established experimentally. This reduces the computational effort for θ_k, which gives more accurate ECM parameter estimations. Equations (10)–(12) will be used in the proposed RLS algorithm, and Equations (5), (7), and (8) will be used to extract the ECM parameters for the purpose of fault diagnosis.

3. Proposed Fault Diagnosis Scheme

In other industrial applications, parameter estimation is a common fault diagnosis method, due to its ability to be implemented online. The method involves the online estimation of the parameters, and the results are compared with a reference model [16]. For real-time identification of ECM parameters, the RLS method is selected because it has low computational demand, fast convergence speed, and can easily be implemented in an embedded system [14]. In this particular case, this method can estimate the model parameters, while adapting to their changes with the degradation and operational conditions of the battery [17]. The resulting estimations are in the form of a time series, for which a change point detection method can be used to diagnose faults [18]. The change point detection method proposed in this paper consists of a WMA filter and a CUSUM control chart.

3.1. Recursive Least Squares Estimation

The RLS algorithm used in this paper employs an optimal forgetting factor to give more weight to recent data, and avoid the saturation phenomenon [19]. The forgetting factor is applied to the parameter vector θ_k. The recursive algorithm of Equation (10) can be represented as follows:

$$K_k = \frac{P_{k-1} \phi_k}{\lambda + \phi_k^T P_{k-1} \phi_k} \tag{13}$$

$$P_k = \frac{P_{k-1} - K_k \phi_k^T P_{k-1}}{\lambda} \tag{14}$$

$$\hat{\theta}_k = \hat{\theta}_{k-1} + K_k (y_k - \hat{\theta}_{k-1}^T \phi_k) \tag{15}$$

where $\hat{\theta}_k$ is the estimated parameter vector θ_k, K_k is the algorithm gain, P_k is the covariance matrix, and λ is the forgetting factor, which will be optimized in the range of [0.95, 1]. The values of θ_0 and P_0 are initially guessed. The schematic diagram for the RLS algorithm is shown in Figure 2.

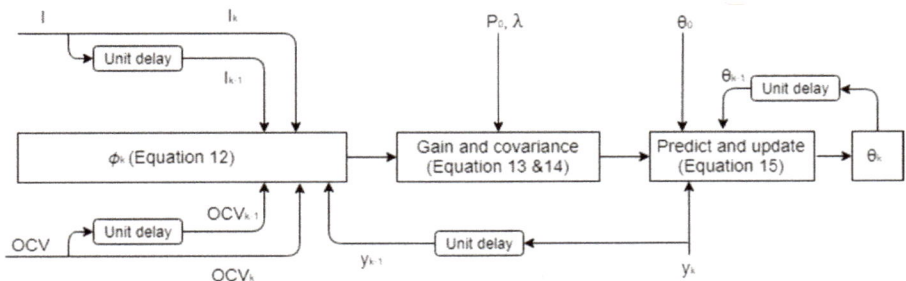

Figure 2. Schematic diagram of the recursive least squares (RLS) algorithm.

3.2. Sensor Faults

A fault is defined as a deviation of at least one property or parameter of the system from the standard condition. Faults are commonly classified as actuator faults, sensor faults, and component/parameter faults. They can affect the control action from the controller, produce measurement errors, or change the input/output properties of the system, which leads to degradation and damage of the system [20]. This paper focuses on sensor faults.

Readings from the sensors in the BMS have an important role in estimating other characteristics of the battery. For instance, the measurements from voltage and current sensors can affect the estimation of SOC. A ±1 mV voltage accuracy system used to calculate SOC in a lithium nickel manganese cobalt oxide (NMC) cell can have a base error of 0.2%. If the same accuracy is used to acquire a lithium iron phosphate (LFP) cell's SOC, then a base error of 5.9% can be expected [21].

The BMS current and voltage sensors used in EVs application can be affected by two types of fault: bias (offset), and gain (scaling) faults. Bias fault is a constant offset from the sensor signal during normal operation. Gain fault happens when the measurement magnitudes are scaled by a factor, while the signal form itself does not change. The faults are considered additive and can be modelled as follows [4]:

$$\tilde{y} = y + f \quad (16)$$

where $\tilde{y}$ is the measured value of current and voltage from the sensors, y is the actual current or voltage, and f is the sensor fault.

3.3. Online Fault Detection Using Weighted Moving Average Filter and Cumulative Sum Control Chart

WMA is a low-pass filter that is used for smoothing fluctuations, such as noise in a time series, to allow for more reliable trend analysis. Additionally, one can use WMA to compute short-term forecasts of time series [22]. The RLS-estimated ECM parameters are time series that contain noise and small fluctuations due to operational conditions (SOC and temperature) and degradation of the cells. A fault, however, is expected to affect the parameters more significantly. Therefore, the difference between WMA-filtered and unfiltered values of the ECM parameters during normal operation of the battery should be considerably smaller than when a fault first occurs. The WMA chosen for the proposed FDI is a two-term WMA to minimize storage requirement. The formula is presented in Equation (17).

$$P_{f,k} = \lambda_{WMA} P_{i,k} + (1 - \lambda_{WMA}) P_{f,k-1} \quad (17)$$

where $P_{f,k}$ is the kth WMA value, $P_{i,k}$ is the kth unfiltered value obtained from RLS (P represents R_0, R_1, and C_1), and λ_{WMA} is the weighting factor. The discrepancy between $P_{f,k}$ and $P_{i,k}$ is characterized by an absolute fractional error term, as shown in Equation (18).

$$e(P_k) = \left|\frac{P_{i,k} - P_{f,k}}{P_{f,k}}\right| \tag{18}$$

The error is monitored using CUSUM, a common change-point detection algorithm, which accumulates deviations of data and signals when the cumulative sum exceeds a certain threshold. The algorithm is outlined in Equation (19) below [23]:

$$S(e(P_k)) = \max\{0, S(e(P_{k-1})) + e(P_k) - (\mu_0 - L\sigma)\} \tag{19}$$

where S is the cumulative sum value, $S(e(P_0)) = 0$; e is the absolute fractional error from Equation (18); μ_0 and σ are the mean and standard deviation of the error population; and L is a specified constant.

In this paper, the λ_{WMA} value from Equation (17) is set to 0.01, since it is more favourable for the filter to obtain a smooth line which can adapt to minor changes over a long period of time, such as noise or degradation effect. In Equation (19), the expected value for μ_0 is 0, and σ is estimated experimentally. During normal operation, the unfiltered values should not deviate from the smooth filtered line, because the amplitude of fluctuation is not significant. When a fault occurs, the unfiltered values would diverge significantly from the smooth filtered series. The CUSUM algorithm detects this divergence by indicating a fault ($F(P_k) = 1$) when $S(e(P_k))$ exceeds an experimentally calibrated threshold J, as shown in Equation (20). When a fault is detected, the BMS will produce an alarm; and appropriate actions, such as replacing the faulty sensor, will be taken to resolve the fault.

$$F(P_k) = \begin{cases} 1 \text{ if } S(e(P_k)) > J \\ 0 \text{ if } S(e(P_k)) < J \end{cases} \tag{20}$$

The method outlined in this section can only be used for fault detection, not fault isolation. The full proposed FDI scheme will be shown in Section 4.5, after determining the effects of different sensor faults on ECM parameters. Since there has not been any work done in literature to determine fault effects on parameters, preliminary experiments will need to be performed to obtain this data before completing the full FDI scheme. The isolation will be based on the response time of the parameters when a certain fault occurs.

4. Effect of Degradation and Faults on ECM Parameters

In order to determine and validate the effect of degradation and faults on the ECM parameters, testing was done on an LFP pouch cell in a laboratory environment. The specifications of the cell at the initial state are listed in Table 1.

Table 1. Lithium iron phosphate (LFP) cell specifications.

Cell Dimension (mm)	7.25 × 160 × 227
Cell Weight (g)	496
Nominal Cell Capacity (Ah)	19
Nominal Cell Voltage (V)	3.3
Voltage Limit (V)	2.0 to 3.65
Operating Temperature (°C)	−30 to 55

4.1. Experimental Setup

The experimental setup consists of a battery test system (Maccor 4200), connected to a testing station and a computer. The full setup is shown in Figure 3. All experiments were carried out at a room temperature of 23 °C. The computer has a software program that controls the battery test system to charge and discharge the cell. The current is assumed to be positive when discharging, and negative when charging. The data was collected at a frequency of 1 Hz, and then stored in the computer. Two test profiles were used: a set of multiple UDDS driving cycles, and a degradation cycle. The UDDS cycle is a velocity profile, and was translated and scaled into a current profile. It was run from the cell SOC of 95% to 20%. The degradation cycle involves charging and discharging multiple times between the extreme limits of the cell to degrade it quickly. Profiles of these cycles are shown in Figure 4. Characterization of the cell was also done through performing the OCV–SOC and capacity tests [24]. The sequence of tests began with cell characterization, then the testing cycle (UDDS and degradation), and all were repeated multiple times.

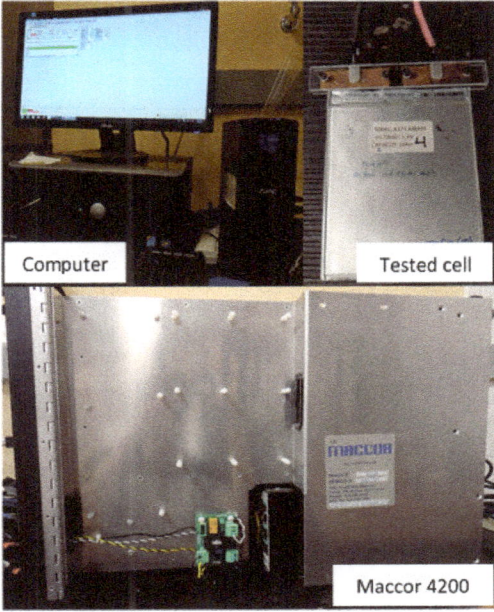

Figure 3. Experimental setup.

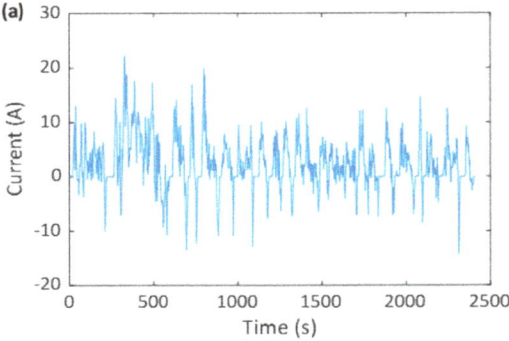

Figure 4. *Cont.*

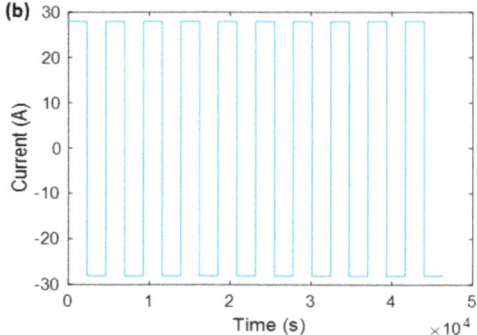

Figure 4. Tested current profiles: (**a**) 1 UDDS cycle; (**b**) 1 degradation cycle.

4.2. Cell Characterization Results

The cell capacity was captured at the beginning of each testing cycle, and it best represents the cell degradation since capacity decreases with degradation [12]. The results are presented in Table 2. The OCV–SOC relationship was also established and a look-up table was built, which was needed to estimate the cell OCV for the RLS algorithm. The OCV–SOC curve was found to change minimally with cell degradation, hence only one curve was used for all cell capacities in the RLS algorithm. The results can be seen in Figure 5.

Table 2. Initial cell capacity for each test cycle.

Cycle	1	2	3	4
Capacity (Ah)	18.26	18.01	17.84	17.66
Cycle	5	6	7	8
Capacity (Ah)	17.32	16.93	16.61	16.47

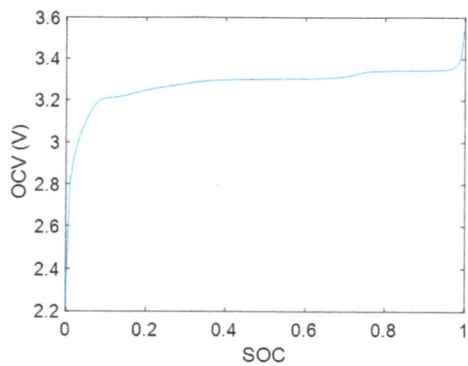

Figure 5. Experimental result for OCV–SOC relationship.

4.3. Effect of Degradation on ECM Parameters

The RLS estimation was used to estimate the ECM parameters for the UDDS driving cycle at different cell capacities. The selected value for λ is 0.9999, as it gives optimal estimation accuracy for the LFP cell tested. Figure 6 shows how degradation affects these parameters. The effect of degradation on R_0 does not show any clear trend. However, it can be clearly seen that R_1 increases, while C_1 decreases, with degradation. This makes sense as the RC pair represents the charge-transfer phenomenon, and degradation can affect the amount of available charge in the battery, which is simply capacity. The changes in these parameters are not significant over a short amount of time, i.e., a few drive cycles, but can be very prominent over the lifetime of the battery. These results confirm that the assumption about the parameters being constant in existing state observer FDI methods, is not valid. Therefore, a reliable FDI scheme should take into consideration the changes in the ECM parameters due to cell degradation.

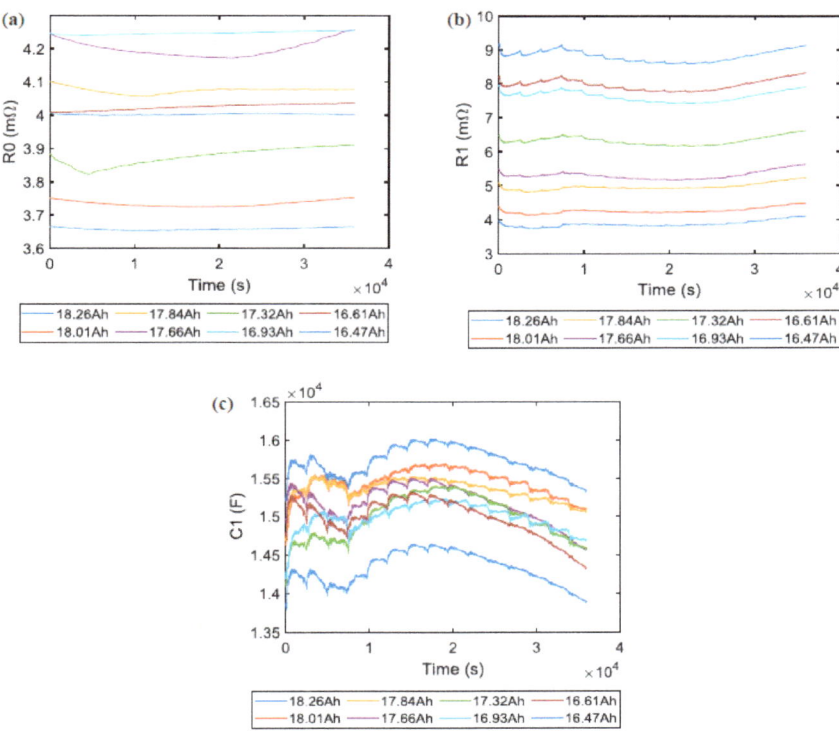

Figure 6. Estimated ECM parameters at various cell capacities. (**a**) R_0 estimation at different cell capacities; (**b**) R_1 estimation at different cell capacities; (**c**) C_1 estimation at different cell capacities.

4.4. Effect of Faults on ECM Parameters

Bias and gain faults were injected into the UDDS driving cycles at various cell capacities, times, and sizes. The effects of the faults were found to be similar across fault types, regardless of the injection time and fault size. The changes in the parameters when the fault is injected can be seen to be more significant, than changes with SOC and temperature [25]. An example is shown in Figure 7, where a voltage gain fault of +10% was injected at the time 30,000 s. When this fault occurs, as shown in Figure 7b,d,f, the parameters diverge away from their original trends. It can also be seen from Figure 7a,c,e that the unfiltered values follow the WMA-filtered line closely during normal operation, while Figure 7b,d,f show that the two lines deviate significantly at the time the fault occurs. This confirms the workability of the proposed change-point detection method using WMA and CUSUM. It is noted that the ECM parameters estimated by RLS require some time to converge. This can be seen at the beginning of Figure 7a–f. Therefore, the proposed FDI scheme would not be able to detect sensor faults for the first hour of battery operation. Considering the long lifespan of Li-ion batteries and the unlikelihood of sensor faults happening within the first hour of operation, it is reasonable to assume there is no fault during the converging period of the RLS algorithm.

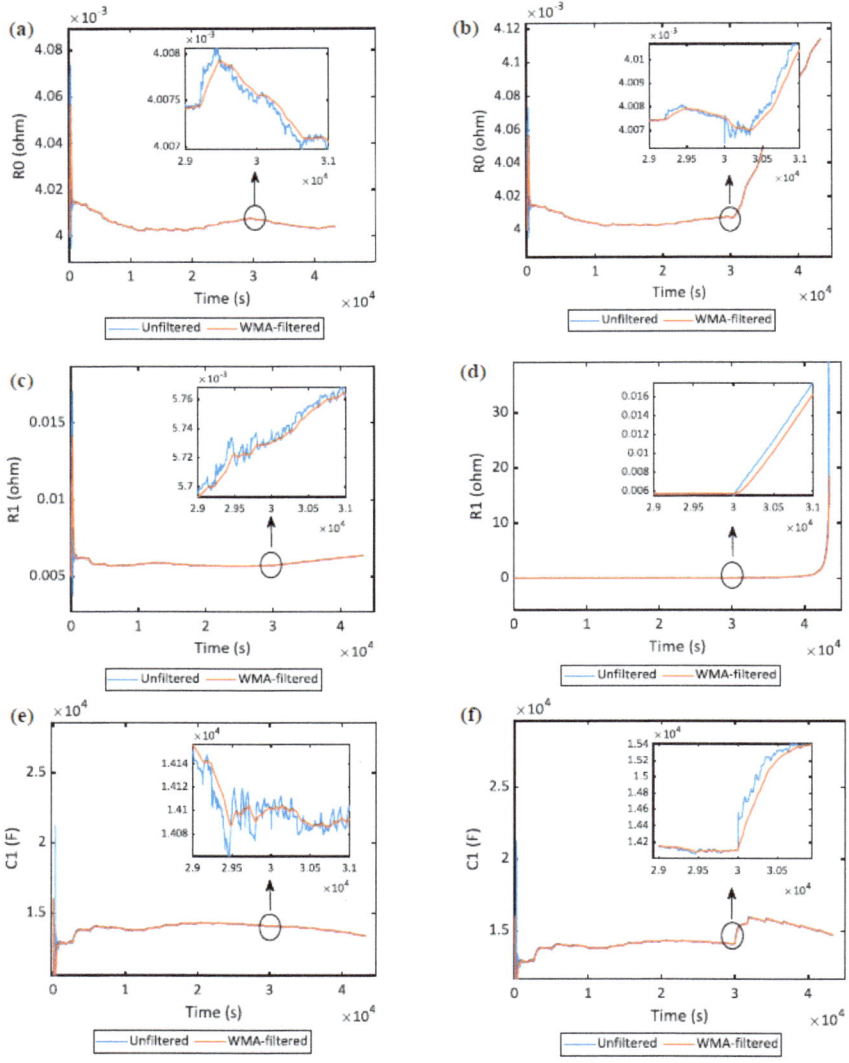

Figure 7. Unfiltered and WMA-filtered ECM parameters during normal operation versus when a fault occurs. (**a**) R_0 during normal operation; (**b**) R_0 when a fault occurs at time 30,000 s; (**c**) R_1 during normal operation; (**d**) R_1 when a fault occurs at time 30,000 s; (**e**) C_1 during normal operation; (**f**) C_1 when a fault occurs at time 30,000 s.

4.5. Isolation of Faults

Through multiple simulations, it was found that R_0 responds the fastest to current sensor faults, while either R_1 or C_1 responds the fastest to voltage sensor faults. From these findings, it is possible to establish a fault isolation schematic to complement the proposed fault detection method. It is uncertain whether these faults would have the same effects on a different type of cell, but this will be focused on and further validated in future studies. For this paper, the FDI scheme will be based on the observations from the tested LFP cell. The full FDI scheme is shown in Figure 8. This scheme will be used to diagnose faults, and validated through simulation in the next section.

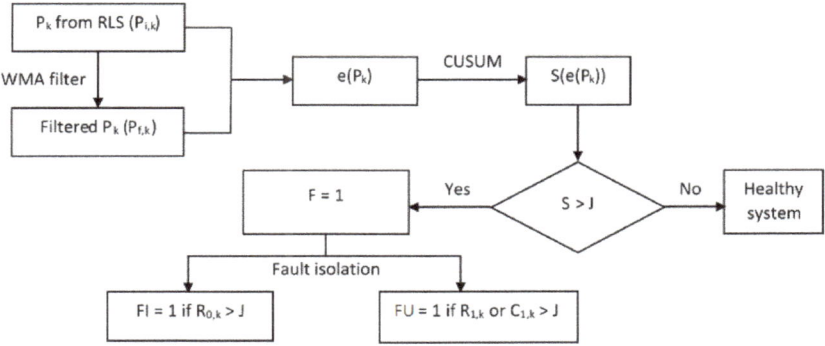

Figure 8. Proposed fault detection and isolation scheme.

5. Diagnostic Implementation and Evaluation

This section shows the validation results of the proposed FDI scheme. The UDDS was selected for use in validation, as it is a realistic daily driving cycle. The experimental runs consisted of multiple UDDS cycles. The experimental setup is described in Section 4.1. The same set of data obtained in Section 4 was also used for the simulations in this section. The simulations started at an SOC of 95% and ended at 20%, and was conducted at various decreasing cell capacities. Faults were injected at random time points. The FDI scheme was validated at all tested capacities to ensure faults can be diagnosed while the cell underwent degradation.

5.1. Voltage Sensor Fault Detection

Multiple voltage sensor faults were injected at different cell capacities in simulation. One specific case will be shown as an example. At a cell capacity of 16.47 Ah, a bias fault of +0.5 V was added to the voltage sensor at the time 30,000 s. The diagnostic results are plotted in Figure 9. Figure 9a,c,e show the deviation between the filtered and unfiltered data. As can be seen, the error increases significantly at the fault injection time. Figure 9b,d,f show the corresponding CUSUM values for the errors. Both the CUSUM values for R_1 and C_1 exceed the threshold at 30,003 s, which is 3 s after the voltage sensor fault occurs. The CUSUM value for R_0 takes longer to respond to the fault, which is expected for voltage sensor faults, and also helps to achieve correct fault isolation. The detected voltage sensor fault signal is plotted in Figure 9g. Table 3 presents results for detection time of the voltage sensor faults of different fault sizes and cell capacities at an injection time of 30,000 s.

Table 3. Summary of detection time for voltage sensor faults at different sizes and cell capacities.

Fault Injected	Capacity (Ah)	18.26	17.84	16.93	16.47
−0.1 V (bias)		19	27	14	14
+0.5 V (bias)	Detection Time (s)	3	3	3	3
+10% (gain)		4	5	4	4

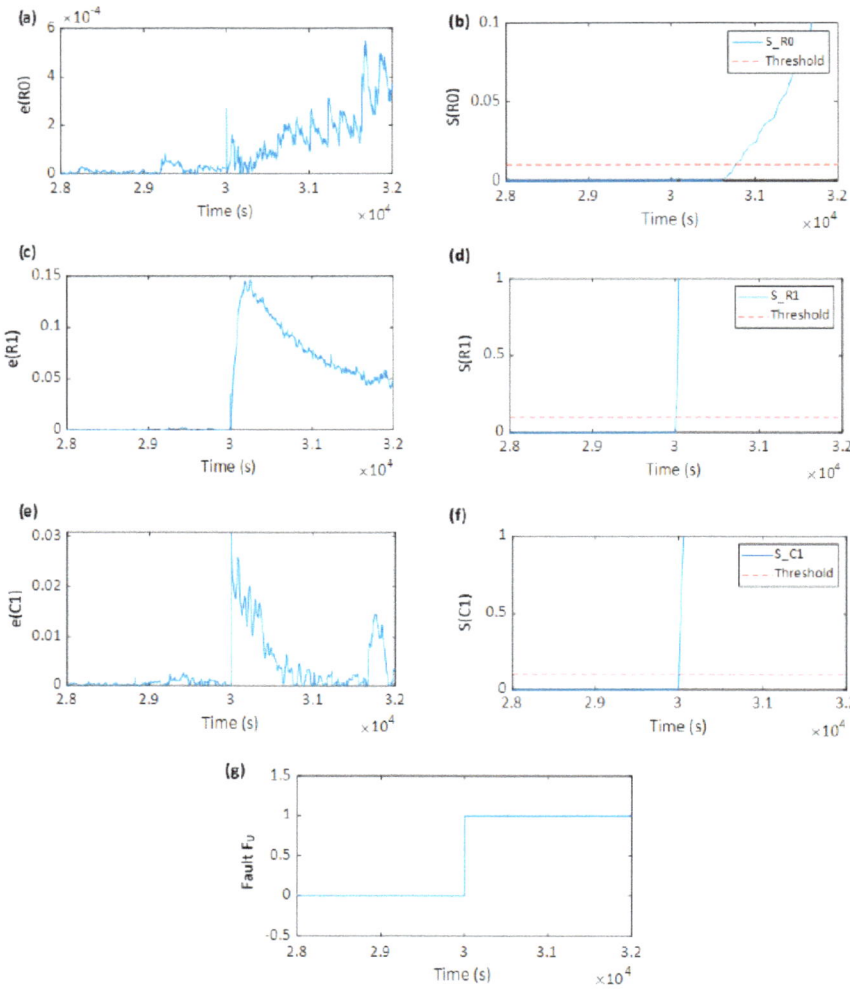

Figure 9. Errors and diagnostic results in the case of voltage sensor fault. (**a**) Error from R_0; (**b**) CUSUM control chart for R_0; (**c**) Error from R_1; (**d**) CUSUM control chart for R_1; (**e**) Error from C_1; (**f**) CUSUM control chart for C_1; (**g**) Isolated voltage sensor fault F_U signal.

5.2. Current Sensor Fault Detection

Similar to the simulation done for voltage sensor fault diagnosis validation, current sensor faults of various sizes were injected at different available cell capacities. The case that will be shown as an example is at a cell capacity of 16.47 Ah, where a gain fault of +10% was injected at the time 30,000 s. The diagnostic results are plotted in Figure 10. The errors were also found to increase at the time of fault injection, as seen in Figure 10a,c,e. Figure 10b,d,f show that the CUSUM values all exceed their respective thresholds after the fault occurs. The CUSUM for the error of R_0 is the fastest to exceed the threshold, at 30,165 s; while the CUSUM values for R_1 and C_1 exceed their thresholds afterward. This indicates a current sensor fault, according to the proposed FDI scheme. Figure 10g shows the detected and isolated current sensor fault signal. The detection time for current sensor faults suffers from a delay, as the CUSUM values take longer to pass their thresholds. Lowering these thresholds should give faster detection time, but risks giving false detection, which is a common trade-off in practice [3]. Table 4 summarizes the results for detection time for the current sensor at an injection time of 30,000 s, with different fault sizes and at different cell capacities.

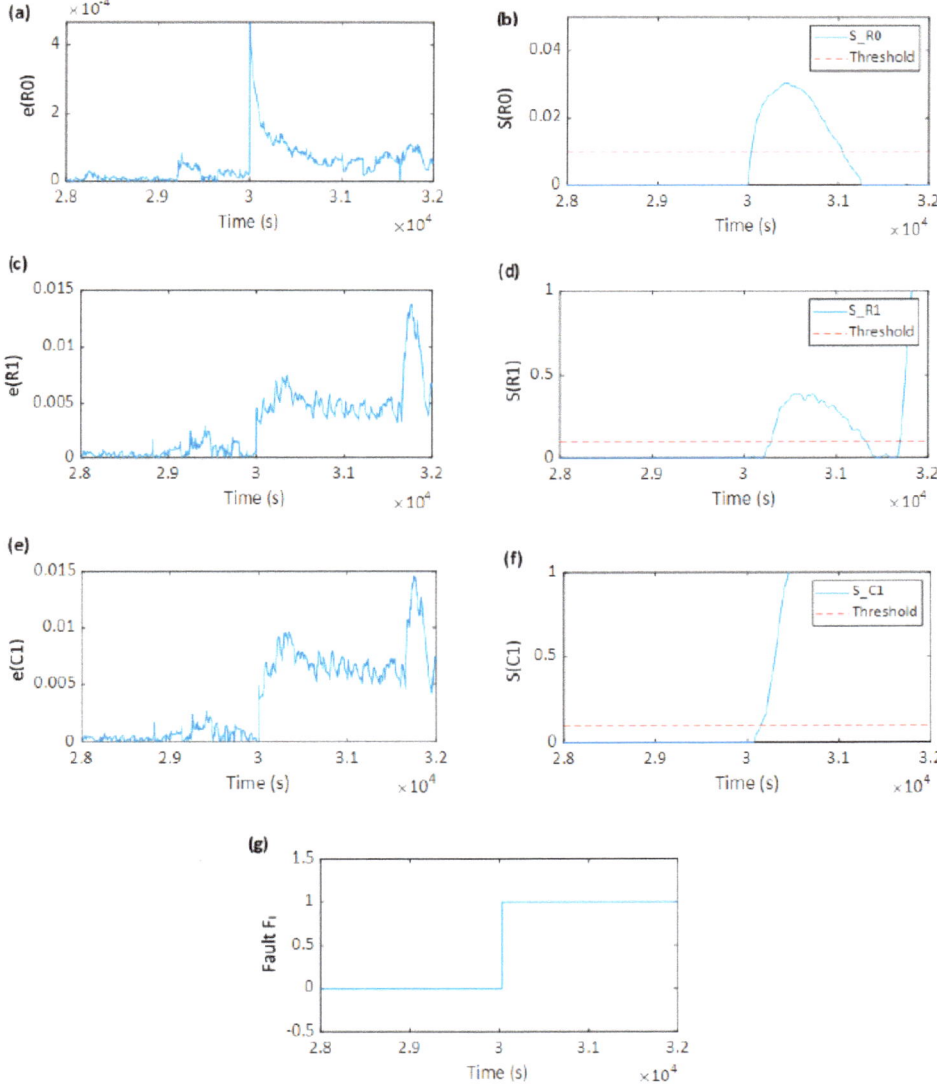

Figure 10. Errors and diagnostic results in the case of current sensor fault. (**a**) Error from R_0; (**b**) CUSUM control chart for R_0; (**c**) Error from R_1; (**d**) CUSUM control chart for R_1; (**e**) Error from C_1; (**f**) CUSUM control chart for C_1; (**g**) Isolated current sensor fault F_I signal.

Table 4. Summary of detection time for current sensor faults at different sizes and cell capacities.

Fault Injected	Capacity (Ah)	18.26	17.84	16.93	16.47
−4 A (bias)	Detection Time (s)	143	190	560	179
+7 A (bias)		45	93	38	37
+10% (gain)		193	181	180	165

For both voltage and current sensors, more simulations were conducted at different injection times, sizes, and capacities to test the validity and effectiveness of the proposed FDI scheme; but it is impossible to show all the results individually, so a summary will be presented. The injection times were set at 10,000 s, 20,000 s, and 30,000 s. It should be noted that faults were not added at the beginning of the runs, due to the proposed FDI scheme's aforementioned inability to detect faults during the converging period of the RLS algorithm, which typically lasts an hour at the start of the battery operation. The considered faults for the voltage sensor are [±0.1 V; ±0.5 V; ±10%], while the considered faults for the current sensor are [±4 A; ±7 A; ±10%]. Approximately 200 runs were simulated. Table 5 shows the results for maximum, minimum, and mean detection time (DT—time from fault occurrence to correct detection of fault), false detection rate (FDR—fraction of tests where fault is detected, but there is no fault) and missed detection rate (MDR—fraction of tests where fault is not detected, but there is a fault). The isolation time depends on the fault size; the larger the fault, the faster the isolation time. It is thus concluded that faults can be detected within a reasonable time using the proposed FDI scheme, with no false detection or missed detection.

Table 5. Summary of the performance evaluation metrics.

	DTmax (s)	DTmin (s)	DTmean (s)	FDR (%)	MDR (%)
Voltage Sensor Fault	127	2	28	0	0
Current Sensor Fault	560	26	172	0	0

6. Conclusions

This paper presented a model-based sensor FDI scheme for a Li-ion cell used in EVs with cell degradation consideration. The scheme uses the RLS algorithm to estimate the ECM parameters in real time, and the WMA filter coupled with CUSUM control chart to detect faults. Experiments and simulations were conducted on an LFP cell in a controlled environment, to verify that ECM parameters are affected by degradation and faults to different degrees; the latter having a more significant effect. It was also found that certain parameters respond faster to specific types of fault, enabling the isolation of faults. Finally, the UDDS driving cycles were used to validate the performance of the proposed FDI scheme. Various injection times, fault sizes, fault types, and cell capacities were considered. The validation results showed that the proposed scheme could detect and isolate voltage sensor faults and current sensor faults for an LFP cell within a reasonable time, with no false or missed detection.

Author Contributions: Conceptualization, M.-K.T. and M.F.; methodology, M.-K.T.; software, M.-K.T.; validation, M.-K.T.; formal analysis, M.-K.T.; investigation, M.-K.T.; resources, M.F.; data curation, M.-K.T.; writing—original draft preparation, M.-K.T.; writing—review and editing, M.-K.T. and M.F.; visualization, M.-K.T.; supervision, M.F.; project administration, M.-K.T.; funding acquisition, M.F. All authors have read and agreed to the published version of the manuscript.

Funding: This research received no external funding.

Acknowledgments: This work was supported by equipment and manpower from the Department of Chemical Engineering at the University of Waterloo.

Conflicts of Interest: The authors declare no conflict of interest.

References

1. Ould Amrouche, S.; Rekioua, D.; Rekioua, T.; Bacha, S. Overview of energy storage in renewable energy systems. *Int. J. Hydrog. Energy* **2016**, *41*, 20914–20927. [CrossRef]
2. Lu, L.; Han, X.; Li, J.; Hua, J.; Ouyang, M. A review on the key issues for lithium-ion battery management in electric vehicles. *J. Power Sources* **2013**, *226*, 272–288. [CrossRef]
3. Liu, Z.; He, H. Sensor fault detection and isolation for a lithium-ion battery pack in electric vehicles using adaptive extended Kalman filter. *Appl. Energy* **2017**, *185*, 2033–2044. [CrossRef]
4. Liu, Z.; He, H. Model-based Sensor Fault Diagnosis of a Lithium-ion Battery in Electric Vehicles. *Energies* **2015**, *8*, 6509–6527. [CrossRef]
5. Wu, C.; Zhu, C.; Ge, Y.; Zhao, Y. A Review on Fault Mechanism and Diagnosis Approach for Li-Ion Batteries. *J. Nanomater.* **2015**, *2015*, 1–9. [CrossRef]
6. Venkatasubramanian, V.; Rengaswamy, R.; Yin, K.; Ka, S.N. A review of process fault detection and diagnosis Part I: Quantitative model-based methods. *Comput. Chem. Eng.* **2003**, *19*, 293–311. [CrossRef]
7. Marcicki, J.; Onori, S.; Rizzoni, G. Nonlinear Fault Detection and Isolation for a Lithium-Ion Battery Management System. In Proceedings of the ASME 2010 Dynamic Systems and Control Conference, 12–15 September 2010; ASMEDC: Cambridge, MA, USA, 2010; Volume 1, pp. 607–614.
8. Chen, W.; Chen, W.-T.; Saif, M.; Li, M.-F.; Wu, H. Simultaneous Fault Isolation and Estimation of Lithium-Ion Batteries via Synthesized Design of Luenberger and Learning Observers. *IEEE Trans. Control Syst. Technol.* **2014**, *22*, 290–298. [CrossRef]
9. Liu, Z.; Ahmed, Q.; Zhang, J.; Rizzoni, G.; He, H. Structural analysis based sensors fault detection and isolation of cylindrical lithium-ion batteries in automotive applications. *Control Eng. Pract.* **2016**, *52*, 46–58. [CrossRef]
10. Dey, S.; Mohon, S.; Pisu, P.; Ayalew, B. Sensor Fault Detection, Isolation, and Estimation in Lithium-Ion Batteries. *IEEE Trans. Control Syst. Technol.* **2016**, *24*, 2141–2149. [CrossRef]
11. Zhang, X.; Lu, J.; Yuan, S.; Yang, J.; Zhou, X. A novel method for identification of lithium-ion battery equivalent circuit model parameters considering electrochemical properties. *J. Power Sources* **2017**, *345*, 21–29. [CrossRef]
12. Yang, G.; Li, J.; Fu, Z.; Guo, L. Adaptive state of charge estimation of Lithium-ion battery based on battery capacity degradation model. *Energy Procedia* **2018**, *152*, 514–519. [CrossRef]
13. Dubarry, M.; Berecibar, M.; Devie, A.; Anseán, D.; Omar, N.; Villarreal, I. State of health battery estimator enabling degradation diagnosis: Model and algorithm description. *J. Power Sources* **2017**, *360*, 59–69. [CrossRef]
14. Fleischer, C.; Waag, W.; Heyn, H.-M.; Sauer, D.U. On-Line adaptive battery impedance parameter and state estimation considering physical principles in reduced order equivalent circuit battery models part 2. Parameter and state estimation. *J. Power Sources* **2014**, *262*, 457–482. [CrossRef]
15. Rahimi-Eichi, H.; Ojha, U.; Baronti, F.; Chow, M.-Y. Battery Management System: An Overview of Its Application in the Smart Grid and Electric Vehicles. *EEE Ind. Electron. Mag.* **2013**, *7*, 4–16. [CrossRef]
16. Che Mid, E.; Dua, V. Model-Based Parameter Estimation for Fault Detection Using Multiparametric Programming. *Ind. Eng. Chem. Res.* **2017**, *56*, 8000–8015. [CrossRef]
17. Duong, V.-H.; Bastawrous, H.A.; Lim, K.; See, K.W.; Zhang, P.; Dou, S.X. Online state of charge and model parameters estimation of the LiFePO 4 battery in electric vehicles using multiple adaptive forgetting factors recursive least-squares. *J. Power Sources* **2015**, *296*, 215–224. [CrossRef]
18. Aminikhanghahi, S.; Cook, D.J. A survey of methods for time series change point detection. *Knowl. Inf. Syst.* **2017**, *51*, 339–367. [CrossRef]
19. He, H.; Zhang, X.; Xiong, R.; Xu, Y.; Guo, H. Online model-based estimation of state-of-charge and open-circuit voltage of lithium-ion batteries in electric vehicles. *Energy* **2012**, *39*, 310–318. [CrossRef]
20. Gao, Z.; Cecati, C.; Ding, S.X. A Survey of Fault Diagnosis and Fault-Tolerant Techniques—Part I: Fault Diagnosis With Model-Based and Signal-Based Approaches. *IEEE Trans. Ind. Electron.* **2015**, *62*, 3757–3767. [CrossRef]
21. Lelie, M.; Braun, T.; Knips, M.; Nordmann, H.; Ringbeck, F.; Zappen, H.; Sauer, D. Battery Management System Hardware Concepts: An Overview. *Appl. Sci.* **2018**, *8*, 534. [CrossRef]

22. Perry, M.B. The Weighted Moving Average Technique. In *Wiley Encyclopedia of Operations Research and Management Science*; John Wiley & Sons, Inc.: Hoboken, NJ, USA, 2011; p. eorms0964. ISBN 978-0-470-40053-1.
23. Jeske, D.R.; Montes De Oca, V.; Bischoff, W.; Marvasti, M. Cusum techniques for timeslot sequences with applications to network surveillance. *Comput. Stat. Data Anal.* **2009**, *53*, 4332–4344. [CrossRef]
24. Christopherson, J.P. *Battery Test Manual for Electric Vehicles*; Idaho National Laboratory: Idaho Falls, ID, USA, 2015.
25. Gomez, J.; Nelson, R.; Kalu, E.E.; Weatherspoon, M.H.; Zheng, J.P. Equivalent circuit model parameters of a high-power Li-ion battery: Thermal and state of charge effects. *J. Power Sources* **2011**, *196*, 4826–4831. [CrossRef]

© 2019 by the authors. Licensee MDPI, Basel, Switzerland. This article is an open access article distributed under the terms and conditions of the Creative Commons Attribution (CC BY) license (http://creativecommons.org/licenses/by/4.0/).

MDPI
St. Alban-Anlage 66
4052 Basel
Switzerland
Tel. +41 61 683 77 34
Fax +41 61 302 89 18
www.mdpi.com

Batteries Editorial Office
E-mail: batteries@mdpi.com
www.mdpi.com/journal/batteries

www.ingramcontent.com/pod-product-compliance
Ingram Content Group UK Ltd.
Pitfield, Milton Keynes, MK11 3LW, UK
UKHW052121230426
12049UKWH00010BA/149